SOCIAL SCIENCE RESEARCH:
PRINCIPLES, METHODS, AND PRACTICES

Anol Bhattacherjee, Ph.D.
University of South Florida
Tampa, Florida, USA
abhatt@usf.edu

Second Edition

Social Science Research: Principles, Methods, and Practices, 2nd edition
By Anol Bhattacherjee
First published 2012

ISBN-13: 978-1475146127
ISBN-10: 1475146124

Preface

This book is designed to introduce doctoral and graduate students to the process of scientific research in the social sciences, business, education, public health, and related disciplines. This book is based on my lecture materials developed over a decade of teaching the doctoral-level class on Research Methods at the University of South Florida. The target audience for this book includes Ph.D. and graduate students, junior researchers, and professors teaching courses on research methods, although senior researchers can also use this book as a handy and compact reference.

The first and most important question potential readers should have about this book is how is it different from other text books on the market? Well, there are four key differences. First, unlike other text books, this book is not just about "research methods" (empirical data collection and analysis) but about the entire "research process" from start to end. Research method is only one phase in that research process, and possibly the easiest and most structured one. Most text books cover research methods in depth, but leave out the more challenging, less structured, and probably more important issues such as theorizing and thinking like a researcher, which are often prerequisites of empirical research. In my experience, most doctoral students become fairly competent at research methods during their Ph.D. years, but struggle to generate interesting or useful research questions or build scientific theories. To address this deficit, I have devoted entire chapters to topics such as "Thinking Like a Researcher" and "Theories in Scientific Research", which are essential skills for a junior researcher.

Second, the book is succinct and compact by design. While writing the book, I decided to focus only on essential concepts, and not fill pages with clutter that can divert the students' attention to less relevant or tangential issues. Most doctoral seminars include a fair complement of readings drawn from the respective discipline. This book is designed to complement those readings by summarizing all important concepts in one compact volume, rather than burden students with a voluminous text on top of their assigned readings.

Third, this book is free in its download version. Not just the current edition but all future editions in perpetuity. The book will also be available in Kindle e-Book, Apple iBook, and on-demand paperback versions at a nominal cost. Many people have asked why I'm giving away something for free when I can make money selling it? Well, not just to stop my students from constantly complaining about the high price of text books, but also because I believe that scientific knowledge should not be constrained by access barriers such as price and availability. Scientific progress can occur only if students and academics around the world have affordable access to the best that science can offer, and this free book is my humble effort to that cause. However, free should not imply "lower quality." Some of the best things in life such as air, water, and sunlight are free. Many of Google's resources are free too, and one can well imagine where we would be in today's Internet age without Google. Some of the most sophisticated software programs available today, like Linux and Apache, are also free, and so is this book.

Fourth, I plan to make local-language versions of this book available in due course of time, and those translated versions will also be free. So far, I have had commitments to translate thus book into Chinese, French, Indonesian, Korean, Portuguese, Spanish versions (which will hopefully be available in 2012), and I'm looking for qualified researchers or professors to translate it into Arabic, German, and other languages where there is sufficient demand for a research text. If you are a prospective translator, please note that there will be no financial gains or royalty for your translation services, because the book must remain free, but I'll gladly include you as a coauthor on the local-language version.

The book is structured into 16 chapters for a 16-week semester. However, professors or instructors can add, drop, stretch, or condense topics to customize the book to the specific needs of their curriculum. For instance, I don't cover Chapters 14 and 15 in my own class, because we have dedicated classes on statistics to cover those materials and more. Instead, I spend two weeks on theories (Chapter 3), one week to discussing and conducting reviews for academic journals (not in the book), and one week for a finals exam. Nevertheless, I felt it necessary to include Chapters 14 and 15 for academic programs that may not have a dedicated class on statistical analysis for research. A sample syllabus that I use for my own class in the business Ph.D. program is provided in the appendix.

Lastly, I plan to continually update this book based on emerging trends in scientific research. If there are any new or interesting content that you wish to see in future editions, please drop me a note, and I will try my best to accommodate them. Comments, criticisms, or corrections to any of the existing content will also be gratefully appreciated.

Anol Bhattacherjee
E-mail: abhatt@usf.edu

Table of Contents

Chapter 1

Science and Scientific Research

What is research? Depending on who you ask, you will likely get very different answers to this seemingly innocuous question. Some people will say that they routinely research different online websites to find the best place to buy goods or services they want. Television news channels supposedly conduct research in the form of viewer polls on topics of public interest such as forthcoming elections or government-funded projects. Undergraduate students research the Internet to find the information they need to complete assigned projects or term papers. Graduate students working on research projects for a professor may see research as collecting or analyzing data related to their project. Businesses and consultants research different potential solutions to remedy organizational problems such as a supply chain bottleneck or to identify customer purchase patterns. However, none of the above can be considered "scientific research" unless: (1) it contributes to a body of science, and (2) it follows the scientific method. This chapter will examine what these terms mean.

Science

What is science? To some, science refers to difficult high school or college-level courses such as physics, chemistry, and biology meant only for the brightest students. To others, science is a craft practiced by scientists in white coats using specialized equipment in their laboratories. Etymologically, the word "science" is derived from the Latin word *scientia* meaning knowledge. **Science** refers to a systematic and organized body of knowledge in any area of inquiry that is acquired using "the scientific method" (the scientific method is described further below). Science can be grouped into two broad categories: natural science and social science. **Natural science** is the science of naturally occurring objects or phenomena, such as light, objects, matter, earth, celestial bodies, or the human body. Natural sciences can be further classified into physical sciences, earth sciences, life sciences, and others. Physical sciences consist of disciplines such as physics (the science of physical objects), chemistry (the science of matter), and astronomy (the science of celestial objects). Earth sciences consist of disciplines such as geology (the science of the earth). Life sciences include disciplines such as biology (the science of human bodies) and botany (the science of plants). In contrast, **social science** is the science of people or collections of people, such as groups, firms, societies, or economies, and their individual or collective behaviors. Social sciences can be classified into disciplines such as psychology (the science of human behaviors), sociology (the science of social groups), and economics (the science of firms, markets, and economies).

The natural sciences are different from the social sciences in several respects. The natural sciences are very precise, accurate, deterministic, and independent of the person making the scientific observations. For instance, a scientific experiment in physics, such as measuring the speed of sound through a certain media or the refractive index of water, should always yield the exact same results, irrespective of the time or place of the experiment, or the person conducting the experiment. If two students conducting the same physics experiment obtain two different values of these physical properties, then it generally means that one or both of those students must be in error. However, the same cannot be said for the social sciences, which tend to be less accurate, deterministic, or unambiguous. For instance, if you measure a person's happiness using a hypothetical instrument, you may find that the same person is more happy or less happy (or sad) on different days and sometimes, at different times on the same day. One's happiness may vary depending on the news that person received that day or on the events that transpired earlier during that day. Furthermore, there is not a single instrument or metric that can accurately measure a person's happiness. Hence, one instrument may calibrate a person as being "more happy" while a second instrument may find that the same person is "less happy" at the same instant in time. In other words, there is a high degree of *measurement error* in the social sciences and there is considerable uncertainty and little agreement on social science policy decisions. For instance, you will not find many disagreements among natural scientists on the speed of light or the speed of the earth around the sun, but you will find numerous disagreements among social scientists on how to solve a social problem such as reduce global terrorism or rescue an economy from a recession. Any student studying the social sciences must be cognizant of and comfortable with handling higher levels of ambiguity, uncertainty, and error that come with such sciences, which merely reflects the high variability of social objects.

Sciences can also be classified based on their purpose. **Basic sciences**, also called pure sciences, are those that explain the most basic objects and forces, relationships between them, and laws governing them. Examples include physics, mathematics, and biology. **Applied sciences**, also called practical sciences, are sciences that apply scientific knowledge from basic sciences in a physical environment. For instance, engineering is an applied science that applies the laws of physics and chemistry for practical applications such as building stronger bridges or fuel efficient combustion engines, while medicine is an applied science that applies the laws of biology for solving human ailments. Both basic and applied sciences are required for human development. However, applied sciences cannot stand on their own right, but instead relies on basic sciences for its progress. Of course, the industry and private enterprises tend to focus more on applied sciences given their practical value, while universities study both basic and applied sciences.

Scientific Knowledge

The purpose of science is to create scientific knowledge. **Scientific knowledge** refers to a generalized body of laws and theories to explain a phenomenon or behavior of interest that are acquired using the scientific method. **Laws** are observed patterns of phenomena or behaviors, while **theories** are systematic explanations of the underlying phenomenon or behavior. For instance, in physics, the Newtonian Laws of Motion describe what happens when an object is in a state of rest or motion (Newton's First Law), what force is needed to move a stationary object or stop a moving object (Newton's Second Law), and what happens when two objects collide (Newton's Third Law). Collectively, the three laws constitute the basis of classical mechanics – a theory of moving objects. Likewise, the theory of optics explains the properties of light and how it behaves in different media, electromagnetic theory explains the properties of electricity and how to generate it, quantum mechanics explains the properties of subatomic particles, and thermodynamics explains the properties of energy and mechanical work. An introductory college level text book in physics will likely contain separate chapters devoted to each of these theories. Similar theories are also available in social sciences. For instance, cognitive dissonance theory in psychology explains how people react when their observations of an event is different from what they expected of that event, general deterrence theory explains why some people engage in improper or criminal behaviors, such as illegally download music or commit software piracy, and the theory of planned behavior explains how people make conscious reasoned choices in their everyday lives.

The goal of scientific research is to discover laws and postulate theories that can explain natural or social phenomena, or in other words, build scientific knowledge. It is important to understand that this knowledge may be imperfect or even quite far from the truth. Sometimes, there may not be a single universal truth, but rather an equilibrium of "multiple truths." We must understand that the theories, upon which scientific knowledge is based, are only explanations of a particular phenomenon, as suggested by a scientist. As such, there may be good or poor explanations, depending on the extent to which those explanations fit well with reality, and consequently, there may be good or poor theories. The progress of science is marked by our progression over time from poorer theories to better theories, through better observations using more accurate instruments and more informed logical reasoning.

We arrive at scientific laws or theories through a process of logic and evidence. Logic (theory) and evidence (observations) are the two, and only two, pillars upon which scientific knowledge is based. In science, theories and observations are interrelated and cannot exist without each other. Theories provide meaning and significance to what we observe, and observations help validate or refine existing theory or construct new theory. Any other means of knowledge acquisition, such as faith or authority cannot be considered science.

Scientific Research

Given that theories and observations are the two pillars of science, scientific research operates at two levels: a theoretical level and an empirical level. The theoretical level is concerned with developing abstract concepts about a natural or social phenomenon and relationships between those concepts (i.e., build "theories"), while the empirical level is concerned with testing the theoretical concepts and relationships to see how well they reflect our observations of reality, with the goal of ultimately building better theories. Over time, a theory becomes more and more refined (i.e., fits the observed reality better), and the science gains maturity. Scientific research involves continually moving back and forth between theory and observations. Both theory and observations are essential components of scientific research. For instance, relying solely on observations for making inferences and ignoring theory is not considered valid scientific research.

Depending on a researcher's training and interest, scientific inquiry may take one of two possible forms: inductive or deductive. In **inductive research**, the goal of a researcher is to infer theoretical concepts and patterns from observed data. In **deductive research**, the goal of the researcher is to test concepts and patterns known from theory using new empirical data. Hence, inductive research is also called *theory-building* research, and deductive research is *theory-testing* research. Note here that the goal of theory-testing is not just to test a theory, but possibly to refine, improve, and extend it. Figure 1.1 depicts the complementary nature of inductive and deductive research. Note that inductive and deductive research are two halves of the research cycle that constantly iterates between theory and observations. You cannot do inductive or deductive research if you are not familiar with both the theory and data components of research. Naturally, a complete researcher is one who can traverse the entire research cycle and can handle both inductive and deductive research.

It is important to understand that theory-building (inductive research) and theory-testing (deductive research) are both critical for the advancement of science. Elegant theories are not valuable if they do not match with reality. Likewise, mountains of data are also useless until they can contribute to the construction to meaningful theories. Rather than viewing these two processes in a circular relationship, as shown in Figure 1.1, perhaps they can be better viewed as a helix, with each iteration between theory and data contributing to better explanations of the phenomenon of interest and better theories. Though both inductive and deductive research are important for the advancement of science, it appears that inductive (theory-building) research is more valuable when there are few prior theories or explanations, while deductive (theory-testing) research is more productive when there are many competing theories of the same phenomenon and researchers are interested in knowing which theory works best and under what circumstances.

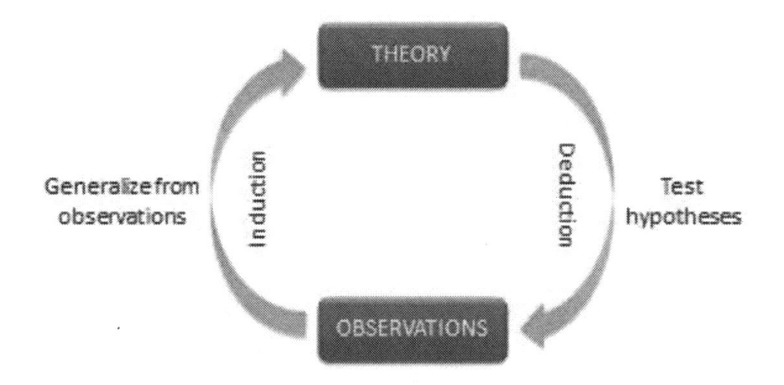

Figure 1.1. The Cycle of Research

Theory building and theory testing are particularly difficult in the social sciences, given the imprecise nature of the theoretical concepts, inadequate tools to measure them, and the presence of many unaccounted factors that can also influence the phenomenon of interest. It is also very difficult to refute theories that do not work. For instance, Karl Marx's theory of communism as an effective means of economic production withstood for decades, before it was finally discredited as being inferior to capitalism in promoting economic growth and social welfare. Erstwhile communist economies like the Soviet Union and China eventually moved toward more capitalistic economies characterized by profit-maximizing private enterprises. However, the recent collapse of the mortgage and financial industries in the United States demonstrates that capitalism also has its flaws and is not as effective in fostering economic growth and social welfare as previously presumed. Unlike theories in the natural sciences, social science theories are rarely perfect, which provides numerous opportunities for researchers to improve those theories or build their own alternative theories.

Conducting scientific research, therefore, requires two sets of skills – theoretical and methodological – needed to operate in the theoretical and empirical levels respectively. Methodological skills ("know-how") are relatively standard, invariant across disciplines, and easily acquired through doctoral programs. However, theoretical skills ("know-what") is considerably harder to master, requires years of observation and reflection, and are tacit skills that cannot be "taught" but rather learned though experience. All of the greatest scientists in the history of mankind, such as Galileo, Newton, Einstein, Neils Bohr, Adam Smith, Charles

Darwin, and Herbert Simon, were master theoreticians, and they are remembered for the theories they postulated that transformed the course of science. Methodological skills are needed to be an ordinary researcher, but theoretical skills are needed to be an extraordinary researcher!

Scientific Method

In the preceding sections, we described science as knowledge acquired through a scientific method. So what exactly is the "scientific method"? **Scientific method** refers to a standardized set of techniques for building scientific knowledge, such as how to make valid observations, how to interpret results, and how to generalize those results. The scientific method allows researchers to independently and impartially test preexisting theories and prior findings, and subject them to open debate, modifications, or enhancements. The scientific method must satisfy four characteristics:

- *Replicability:* Others should be able to independently replicate or repeat a scientific study and obtain similar, if not identical, results.

- *Precision:* Theoretical concepts, which are often hard to measure, must be defined with such precision that others can use those definitions to measure those concepts and test that theory.

- *Falsifiability:* A theory must be stated in a way that it can be disproven. Theories that cannot be tested or falsified are not scientific theories and any such knowledge is not scientific knowledge. A theory that is specified in imprecise terms or whose concepts are not accurately measurable cannot be tested, and is therefore not scientific. Sigmund Freud's ideas on psychoanalysis fall into this category and is therefore not considered a "theory", even though psychoanalysis may have practical utility in treating certain types of ailments.

- *Parsimony:* When there are multiple explanations of a phenomenon, scientists must always accept the simplest or logically most economical explanation. This concept is called parsimony or "Occam's razor." Parsimony prevents scientists from pursuing overly complex or outlandish theories with endless number of concepts and relationships that may explain a little bit of everything but nothing in particular.

Any branch of inquiry that does not allow the scientific method to test its basic laws or theories cannot be called "science." For instance, theology (the study of religion) is not science because theological ideas (such as the presence of God) cannot be tested by independent observers using a replicable, precise, falsifiable, and parsimonious method. Similarly, arts, music, literature, humanities, and law are also not considered science, even though they are creative and worthwhile endeavors in their own right.

The scientific method, as applied to social sciences, includes a variety of research approaches, tools, and techniques, such as qualitative and quantitative data, statistical analysis, experiments, field surveys, case research, and so forth. Most of this book is devoted to learning about these different methods. However, recognize that the scientific method operates primarily at the empirical level of research, i.e., how to make observations and analyze and interpret these observations. Very little of this method is directly pertinent to the theoretical level, which is really the more challenging part of scientific research.

Types of Scientific Research

Depending on the purpose of research, scientific research projects can be grouped into three types: exploratory, descriptive, and explanatory. **Exploratory research** is often conducted in new areas of inquiry, where the goals of the research are: (1) to scope out the magnitude or extent of a particular phenomenon, problem, or behavior, (2) to generate some initial ideas (or "hunches") about that phenomenon, or (3) to test the feasibility of undertaking a more extensive study regarding that phenomenon. For instance, if the citizens of a country are generally dissatisfied with governmental policies regarding during an economic recession, exploratory research may be directed at measuring the extent of citizens' dissatisfaction, understanding how such dissatisfaction is manifested, such as the frequency of public protests, and the presumed causes of such dissatisfaction, such as ineffective government policies in dealing with inflation, interest rates, unemployment, or higher taxes. Such research may include examination of publicly reported figures, such as estimates of economic indicators, such as gross domestic product (GDP), unemployment, and consumer price index, as archived by third-party sources, obtained through interviews of experts, eminent economists, or key government officials, and/or derived from studying historical examples of dealing with similar problems. This research may not lead to a very accurate understanding of the target problem, but may be worthwhile in scoping out the nature and extent of the problem and serve as a useful precursor to more in-depth research.

Descriptive research is directed at making careful observations and detailed documentation of a phenomenon of interest. These observations must be based on the scientific method (i.e., must be replicable, precise, etc.), and therefore, are more reliable than casual observations by untrained people. Examples of descriptive research are tabulation of demographic statistics by the United States Census Bureau or employment statistics by the Bureau of Labor, who use the same or similar instruments for estimating employment by sector or population growth by ethnicity over multiple employment surveys or censuses. If any changes are made to the measuring instruments, estimates are provided with and without the changed instrumentation to allow the readers to make a fair before-and-after comparison regarding population or employment trends. Other descriptive research may include chronicling ethnographic reports of gang activities among adolescent youth in urban populations, the persistence or evolution of religious, cultural, or ethnic practices in select communities, and the role of technologies such as Twitter and instant messaging in the spread of democracy movements in Middle Eastern countries.

Explanatory research seeks explanations of observed phenomena, problems, or behaviors. While descriptive research examines the what, where, and when of a phenomenon, explanatory research seeks answers to why and how types of questions. It attempts to "connect the dots" in research, by identifying causal factors and outcomes of the target phenomenon. Examples include understanding the reasons behind adolescent crime or gang violence, with the goal of prescribing strategies to overcome such societal ailments. Most academic or doctoral research belongs to the explanation category, though some amount of exploratory and/or descriptive research may also be needed during initial phases of academic research. Seeking explanations for observed events requires strong theoretical and interpretation skills, along with intuition, insights, and personal experience. Those who can do it well are also the most prized scientists in their disciplines.

History of Scientific Thought

Before closing this chapter, it may be interesting to go back in history and see how science has evolved over time and identify the key scientific minds in this evolution. Although instances of scientific progress have been documented over many centuries, the terms "science," "scientists," and the "scientific method" were coined only in the 19th century. Prior to this time, science was viewed as a part of philosophy, and coexisted with other branches of philosophy such as logic, metaphysics, ethics, and aesthetics, although the boundaries between some of these branches were blurred.

In the earliest days of human inquiry, knowledge was usually recognized in terms of theological precepts based on faith. This was challenged by Greek philosophers such as Plato, Aristotle, and Socrates during the 3rd century BC, who suggested that the fundamental nature of being and the world can be understood more accurately through a process of systematic logical reasoning called **rationalism**. In particular, Aristotle's classic work *Metaphysics* (literally meaning "beyond physical [existence]") separated *theology* (the study of Gods) from *ontology* (the study of being and existence) and *universal science* (the study of first principles, upon which logic is based). Rationalism (not to be confused with "rationality") views reason as the source of knowledge or justification, and suggests that the criterion of truth is not sensory but rather intellectual and deductive, often derived from a set of first principles or axioms (such as Aristotle's "law of non-contradiction").

The next major shift in scientific thought occurred during the 16th century, when British philosopher Francis Bacon (1561-1626) suggested that knowledge can only be derived from observations in the real world. Based on this premise, Bacon emphasized knowledge acquisition as an empirical activity (rather than as a reasoning activity), and developed **empiricism** as an influential branch of philosophy. Bacon's works led to the popularization of inductive methods of scientific inquiry, the development of the "scientific method" (originally called the "Baconian method"), consisting of systematic observation, measurement, and experimentation, and may have even sowed the seeds of atheism or the rejection of theological precepts as "unobservable."

Empiricism continued to clash with rationalism throughout the Middle Ages, as philosophers sought the most effective way of gaining valid knowledge. French philosopher Rene Descartes sided with the rationalists, while British philosophers John Locke and David Hume sided with the empiricists. Other scientists, such as Galileo Galilei and Sir Issac Newton, attempted to fuse the two ideas into **natural philosophy** (the philosophy of nature), to focus specifically on understanding nature and the physical universe, which is considered to be the precursor of the natural sciences. Galileo (1564-1642) was perhaps the first to state that the laws of nature are mathematical, and contributed to the field of astronomy through an innovative combination of experimentation and mathematics.

In the 18th century, German philosopher Immanuel Kant sought to resolve the dispute between empiricism and rationalism in his book *Critique of Pure Reason*, by arguing that experience is purely subjective and processing them using pure reason without first delving into the subjective nature of experiences will lead to theoretical illusions. Kant's ideas led to the development of **German idealism**, which inspired later development of interpretive techniques such as phenomenology, hermeneutics, and critical social theory.

At about the same time, French philosopher Auguste Comte (1798–1857), founder of the discipline of sociology, attempted to blend rationalism and empiricism in a new doctrine called **positivism**. He suggested that theory and observations have circular dependence on each other. While theories may be created via reasoning, they are only authentic if they can be verified through observations. The emphasis on verification started the separation of modern science from philosophy and metaphysics and further development of the "scientific method" as the primary means of validating scientific claims. Comte's ideas were expanded by Emile Durkheim in his development of sociological positivism (positivism as a foundation for social research) and Ludwig Wittgenstein in logical positivism.

In the early 20th century, strong accounts of positivism were rejected by interpretive sociologists (antipositivists) belonging to the German idealism school of thought. Positivism was typically equated with quantitative research methods such as experiments and surveys and without any explicit philosophical commitments, while **antipositivism** employed qualitative methods such as unstructured interviews and participant observation. Even practitioners of positivism, such as American sociologist Paul Lazarsfield who pioneered large-scale survey research and statistical techniques for analyzing survey data, acknowledged potential problems of observer bias and structural limitations in positivist inquiry. In response, antipositivists emphasized that social actions must be studied though interpretive means based upon an understanding the meaning and purpose that individuals attach to their personal actions, which inspired Georg Simmel's work on symbolic interactionism, Max Weber's work on ideal types, and Edmund Husserl's work on phenomenology.

In the mid-to-late 20th century, both positivist and antipositivist schools of thought were subjected to criticisms and modifications. British philosopher Sir Karl Popper suggested that human knowledge is based not on unchallengeable, rock solid foundations, but rather on a set of tentative conjectures that can never be proven conclusively, but only disproven. Empirical evidence is the basis for disproving these conjectures or "theories." This metatheoretical stance, called **postpositivism** (or postempiricism), amends positivism by suggesting that it is impossible to verify the truth although it is possible to reject false beliefs, though it retains the positivist notion of an objective truth and its emphasis on the scientific method.

Likewise, antipositivists have also been criticized for trying only to understand society but not critiquing and changing society for the better. The roots of this thought lie in *Das Capital*, written by German philosophers Karl Marx and Friedrich Engels, which critiqued capitalistic societies as being social inequitable and inefficient, and recommended resolving this inequity through class conflict and proletarian revolutions. Marxism inspired social revolutions in countries such as Germany, Italy, Russia, and China, but generally failed to accomplish the social equality that it aspired. **Critical research** (also called critical theory) propounded by Max Horkheimer and Jurgen Habermas in the 20th century, retains similar ideas of critiquing and resolving social inequality, and adds that people can and should consciously act to change their social and economic circumstances, although their ability to do so is constrained by various forms of social, cultural and political domination. Critical research attempts to uncover and critique the restrictive and alienating conditions of the status quo by analyzing the oppositions, conflicts and contradictions in contemporary society, and seeks to eliminate the causes of alienation and domination (i.e., emancipate the oppressed class). More on these different research philosophies and approaches will be covered in future chapters of this book.

Chapter 2

Thinking Like a Researcher

Conducting good research requires first retraining your brain to think like a researcher. This requires visualizing the abstract from actual observations, mentally "connecting the dots" to identify hidden concepts and patterns, and synthesizing those patterns into generalizable laws and theories that apply to other contexts beyond the domain of the initial observations. Research involves constantly moving back and forth from an empirical plane where observations are conducted to a theoretical plane where these observations are abstracted into generalizable laws and theories. This is a skill that takes many years to develop, is not something that is taught in graduate or doctoral programs or acquired in industry training, and is by far the biggest deficit amongst Ph.D. students. Some of the mental abstractions needed to think like a researcher include unit of analysis, constructs, hypotheses, operationalization, theories, models, induction, deduction, and so forth, which we will examine in this chapter.

Unit of Analysis

One of the first decisions in any social science research is the unit of analysis of a scientific study. The **unit of analysis** refers to the person, collective, or object that is the target of the investigation. Typical unit of analysis include individuals, groups, organizations, countries, technologies, objects, and such. For instance, if we are interested in studying people's shopping behavior, their learning outcomes, or their attitudes to new technologies, then the unit of analysis is the *individual*. If we want to study characteristics of street gangs or teamwork in organizations, then the unit of analysis is the *group*. If the goal of research is to understand how firms can improve profitability or make good executive decisions, then the unit of analysis is the *firm*. In this case, even though decisions are made by individuals in these firms, these individuals are presumed to represent their firm's decision rather than their personal decisions. If research is directed at understanding differences in national cultures, then the unit of analysis becomes a *country*. Even inanimate objects can serve as units of analysis. For instance, if a researcher is interested in understanding how to make web pages more attractive to its users, then the unit of analysis is a *web page* (and not users). If we wish to study how knowledge transfer occurs between two firms, then our unit of analysis becomes the *dyad* (the combination of firms that is sending and receiving knowledge).

Understanding the units of analysis can sometimes be fairly complex. For instance, if we wish to study why certain neighborhoods have high crime rates, then our unit of analysis becomes the *neighborhood*, and not crimes or criminals committing such crimes. This is because the object of our inquiry is the neighborhood and not criminals. However, if we wish to compare different types of crimes in different neighborhoods, such as homicide, robbery, assault, and so forth, our unit of analysis becomes the *crime*. If we wish to study why criminals engage in illegal activities, then the unit of analysis becomes the *individual* (i.e., the criminal). Like, if we want to study why some innovations are more successful than others, then our unit of analysis is an *innovation*. However, if we wish to study how some organizations innovate more consistently than others, then the unit of analysis is the *organization*. Hence, two related research questions within the same research study may have two entirely different units of analysis.

Understanding the unit of analysis is important because it shapes what type of data you should collect for your study and who you collect it from. If your unit of analysis is a web page, you should be collecting data about web pages from actual web pages, and not surveying people about how they use web pages. If your unit of analysis is the organization, then you should be measuring organizational-level variables such as organizational size, revenues, hierarchy, or absorptive capacity. This data may come from a variety of sources such as financial records or surveys of Chief Executive Officers (CEO), who are presumed to be representing their organization (rather than themselves). Some variables such as CEO pay may seem like individual level variables, but in fact, it can also be an organizational level variable because each organization has only one CEO pay at any time. Sometimes, it is possible to collect data from a lower level of analysis and aggregate that data to a higher level of analysis. For instance, in order to study teamwork in organizations, you can survey individual team members in different organizational teams, and average their individual scores to create a composite team-level score for team-level variables like cohesion and conflict. We will examine the notion of "variables" in greater depth in the next section.

Concepts, Constructs, and Variables

We discussed in Chapter 1 that although research can be exploratory, descriptive, or explanatory, most scientific research tend to be of the explanatory type in that they search for potential explanations of observed natural or social phenomena. Explanations require development of **concepts** or generalizable properties or characteristics associated with objects, events, or people. While objects such as a person, a firm, or a car are not concepts, their specific characteristics or behavior such as a person's attitude toward immigrants, a firm's capacity for innovation, and a car's weight can be viewed as concepts.

Knowingly or unknowingly, we use different kinds of concepts in our everyday conversations. Some of these concepts have been developed over time through our shared language. Sometimes, we borrow concepts from other disciplines or languages to explain a phenomenon of interest. For instance, the idea of *gravitation* borrowed from physics can be used in business to describe why people tend to "gravitate" to their preferred shopping destinations. Likewise, the concept of *distance* can be used to explain the degree of social separation between two otherwise collocated individuals. Sometimes, we create our own concepts to describe a unique characteristic not described in prior research. For instance, *technostress* is a new concept referring to the mental stress one may face when asked to learn a new technology.

Concepts may also have progressive levels of abstraction. Some concepts such as a person's *weight* are precise and objective, while other concepts such as a person's *personality* may be more abstract and difficult to visualize. A **construct** is an abstract concept that is specifically chosen (or "created") to explain a given phenomenon. A construct may be a simple concept, such as a person's *weight*, or a combination of a set of related concepts such as a person's *communication skill*, which may consist of several underlying concepts such as the person's *vocabulary, syntax*, and *spelling*. The former instance (weight) is a **unidimensional construct**, while the latter (communication skill) is a **multi-dimensional construct** (i.e., it consists of multiple underlying concepts). The distinction between constructs and concepts are clearer in multi-dimensional constructs, where the higher order abstraction is called a construct and the lower order abstractions are called concepts. However, this distinction tends to blur in the case of unidimensional constructs.

Constructs used for scientific research must have precise and clear definitions that others can use to understand exactly what it means and what it does not mean. For instance, a seemingly simple construct such as *income* may refer to monthly or annual income, before-tax or after-tax income, and personal or family income, and is therefore neither precise nor clear. There are two types of definitions: dictionary definitions and operational definitions. In the more familiar dictionary definition, a construct is often defined in terms of a synonym. For instance, attitude may be defined as a disposition, a feeling, or an affect, and affect in turn is defined as an attitude. Such definitions of a circular nature are not particularly useful in scientific research for elaborating the meaning and content of that construct. Scientific research requires **operational definitions** that define constructs in terms of how they will be empirically measured. For instance, the operational definition of a construct such as *temperature* must specify whether we plan to measure temperature in Celsius, Fahrenheit, or Kelvin scale. A construct such as *income* should be defined in terms of whether we are interested in monthly or annual income, before-tax or after-tax income, and personal or family income. One can imagine that constructs such as *learning*, *personality*, and *intelligence* can be quite hard to define operationally.

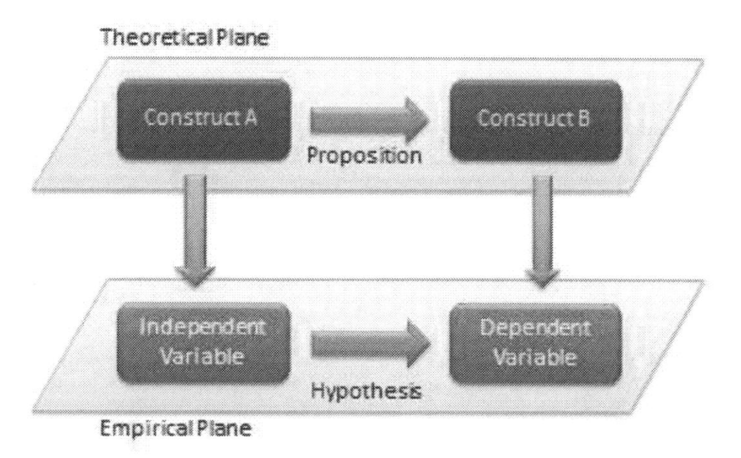

Figure 2.1. The theoretical and empirical planes of research

A term frequently associated with, and sometimes used interchangeably with, a construct is a variable. Etymologically speaking, a variable is a quantity that can vary (e.g., from low to high, negative to positive, etc.), in contrast to constants that do not vary (i.e., remain constant). However, in scientific research, a **variable** is a measurable representation of an abstract construct. As abstract entities, constructs are not directly measurable, and hence, we look for proxy measures called variables. For instance, a person's *intelligence* is often measured as his or her *IQ (intelligence quotient) score*, which is an index generated from an analytical and pattern-matching test administered to people. In this case, *intelligence* is a construct, and *IQ score* is a variable that measures the intelligence construct. Whether IQ scores truly measures one's intelligence is anyone's guess (though many believe that they do), and depending on whether how well it measures intelligence, the IQ score may be a good or a poor measure of the intelligence construct. As shown in Figure 2.1, scientific research proceeds along two planes: a theoretical plane and an empirical plane. Constructs are conceptualized at the theoretical (abstract) plane, while variables are operationalized and measured at the empirical (observational) plane. Thinking like a researcher implies the ability to move back and forth between these two planes.

Depending on their intended use, variables may be classified as independent, dependent, moderating, mediating, or control variables. Variables that explain other variables are called **independent variables**, those that are explained by other variables are **dependent variables**, those that are explained by independent variables while also explaining dependent variables are **mediating variables** (or intermediate variables), and those that influence the relationship between independent and dependent variables are called **moderating variables**. As an example, if we state that higher intelligence causes improved learning among students, then intelligence is an independent variable and learning is a dependent variable. There may be other extraneous variables that are not pertinent to explaining a given dependent variable, but may have some impact on the dependent variable. These variables must be controlled for in a scientific study, and are therefore called **control variables**.

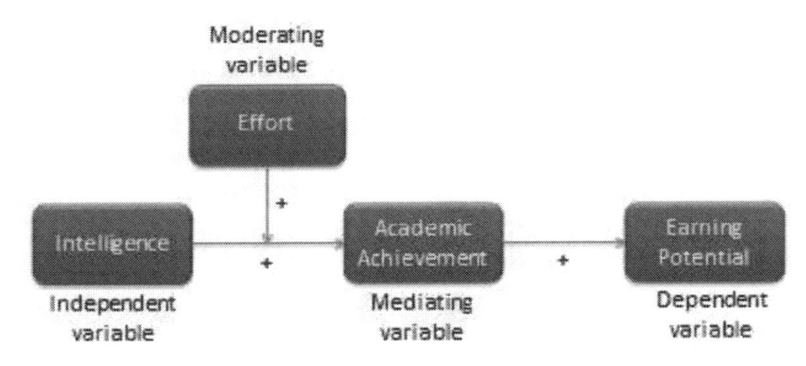

Figure 2.2. A nomological network of constructs

To understand the differences between these different variable types, consider the example shown in Figure 2.2. If we believe that intelligence influences (or explains) students' academic achievement, then a measure of intelligence such as an *IQ score* is an independent variable, while a measure of academic success such as *grade point average* is a dependent variable. If we believe that the effect of intelligence on academic achievement also depends on the effort invested by the student in the learning process (i.e., between two equally intelligent students, the student who puts is more effort achieves higher academic achievement than one who puts in less effort), then *effort* becomes a moderating variable. Incidentally, one may also view effort as an independent variable and intelligence as a moderating variable. If academic achievement is viewed as an intermediate step to higher earning potential, then *earning potential* becomes the dependent variable for the independent variable *academic achievement*, and academic achievement becomes the mediating variable in the relationship between intelligence and earning potential. Hence, variable are defined as an independent, dependent, moderating, or mediating variable based on their nature of association with each other. The overall network of relationships between a set of related constructs is called a **nomological network** (see Figure 2.2). Thinking like a researcher requires not only being able to abstract constructs from observations, but also being able to mentally visualize a nomological network linking these abstract constructs.

Propositions and Hypotheses

Figure 2.2 shows how theoretical constructs such as intelligence, effort, academic achievement, and earning potential are related to each other in a nomological network. Each of these relationships is called a proposition. In seeking explanations to a given phenomenon or behavior, it is not adequate just to identify key concepts and constructs underlying the target phenomenon or behavior. We must also identify and state patterns of relationships between these constructs. Such patterns of relationships are called propositions. A **proposition** is a tentative and conjectural relationship between constructs that is stated in a declarative form. An example of a proposition is: "An increase in student intelligence causes an increase in their academic achievement." This declarative statement does not have to be true, but must be empirically testable using data, so that we can judge whether it is true or false. Propositions are generally derived based on logic (deduction) or empirical observations (induction).

Because propositions are associations between abstract constructs, they cannot be tested directly. Instead, they are tested indirectly by examining the relationship between corresponding measures (variables) of those constructs. The empirical formulation of propositions, stated as relationships between variables, is called **hypotheses** (see Figure 2.1). Since IQ scores and grade point average are operational measures of intelligence and academic achievement respectively, the above proposition can be specified in form of the hypothesis: "An increase in students' IQ score causes an increase in their grade point average." Propositions are specified in the theoretical plane, while hypotheses are specified in the empirical plane. Hence, hypotheses are empirically testable using observed data, and may be rejected if not supported by empirical observations. Of course, the goal of hypothesis testing is to infer whether the corresponding proposition is valid.

Hypotheses can be strong or weak. "Students' IQ scores are related to their academic achievement" is an example of a weak hypothesis, since it indicates neither the directionality of the hypothesis (i.e., whether the relationship is positive or negative), nor its causality (i.e., whether intelligence causes academic achievement or academic achievement causes intelligence). A stronger hypothesis is "students' IQ scores are *positively* related to their academic achievement", which indicates the directionality but not the causality. A still better hypothesis is "students' IQ scores have positive effects on their academic achievement", which specifies both the directionality and the causality (i.e., intelligence causes academic achievement, and not the reverse). The signs in Figure 2.2 indicate the directionality of the respective hypotheses.

Also note that scientific hypotheses should clearly specify independent and dependent variables. In the hypothesis, "students' IQ scores have positive effects on their academic achievement," it is clear that intelligence is the independent variable (the "cause") and academic achievement is the dependent variable (the "effect"). Further, it is also clear that this hypothesis can be evaluated as either true (if higher intelligence leads to higher academic achievement) or false (if higher intelligence has no effect on or leads to lower academic achievement). Later on in this book, we will examine how to empirically test such cause-effect relationships. Statements such as "students are generally intelligent" or "all students can achieve academic success" are not scientific hypotheses because they do not specify independent and dependent variables, nor do they specify a directional relationship that can be evaluated as true or false.

Theories and Models

A **theory** is a set of systematically interrelated constructs and propositions intended to explain and predict a phenomenon or behavior of interest, within certain boundary conditions and assumptions. Essentially, a theory is a systemic collection of related theoretical propositions. While propositions generally connect two or three constructs, theories represent a *system* of multiple constructs and propositions. Hence, theories can be substantially more complex and abstract and of a larger scope than propositions or hypotheses.

I must note here that people not familiar with scientific research often view a theory as a *speculation* or the opposite of *fact*. For instance, people often say that teachers need to be less theoretical and more practical or factual in their classroom teaching. However, practice or fact are not opposites of theory, but in a scientific sense, are essential components needed to test the validity of a theory. A good scientific theory should be well supported using observed facts and should also have practical value, while a poorly defined theory tends to be lacking in these dimensions. Famous organizational research Kurt Lewin once said, "Theory without practice is sterile; practice without theory is blind." Hence, both theory and facts (or practice) are essential for scientific research.

Theories provide explanations of social or natural phenomenon. As emphasized in Chapter 1, these explanations may be good or poor. Hence, there may be good or poor theories. Chapter 3 describes some criteria that can be used to evaluate how good a theory really is. Nevertheless, it is important for researchers to understand that theory is not "truth," there is nothing sacrosanct about any theory, and theories should not be accepted just because they were proposed by someone. In the course of scientific progress, poorer theories are eventually replaced by better theories with higher explanatory power. The essential challenge for researchers is to build better and more comprehensive theories that can explain a target phenomenon better than prior theories.

A term often used in conjunction with theory is a model. A **model** is a representation of all or part of a system that is constructed to study that system (e.g., how the system works or what triggers the system). While a theory tries to explain a phenomenon, a model tries to represent a phenomenon. Models are often used by decision makers to make important decisions based on a given set of inputs. For instance, marketing managers may use models to decide how much money to spend on advertising for different product lines based on parameters such as prior year's advertising expenses, sales, market growth, and competing products. Likewise, weather forecasters can use models to predict future weather patterns based on parameters such as wind speeds, wind direction, temperature, and humidity. While these models are useful, they may not necessarily explain advertising expenditure or weather forecasts. Models may be of different kinds, such as mathematical models, network models, and path models. Models can also be descriptive, predictive, or normative. Descriptive models are frequently used for representing complex systems, for visualizing variables and relationships in such systems. An advertising expenditure model may be a descriptive model. Predictive models (e.g., a regression model) allow forecast of future events. Weather forecasting models are predictive models. Normative models are used to guide our activities along commonly accepted norms or practices. Models may also be static if it represents the state of a system at one point in time, or dynamic, if it represents a system's evolution over time.

The process of theory or model development may involve inductive and deductive reasoning. Recall from Chapter 1 that **deduction** is the process of drawing conclusions about a phenomenon or behavior based on theoretical or logical reasons and an initial set of premises. As an example, if a certain bank enforces a strict code of ethics for its employees (Premise 1) and Jamie is an employee at that bank (Premise 2), then Jamie can be trusted to follow ethical practices (Conclusion). In deduction, the conclusions must be true if the initial premises and reasons are correct.

In contrast, **induction** is the process of drawing conclusions based on facts or observed evidence. For instance, if a firm spent a lot of money on a promotional campaign (Observation 1), but the sales did not increase (Observation 2), then possibly the promotion campaign was poorly executed (Conclusion). However, there may be rival explanations for poor sales, such as economic recession or the emergence of a competing product or brand or perhaps a supply chain problem. Inductive conclusions are therefore only a hypothesis, and may be disproven. Deductive conclusions generally tend to be stronger than inductive conclusions, but a deductive conclusion based on an incorrect premise is also incorrect.

As shown in Figure 2.3, inductive and deductive reasoning go hand in hand in theory and model building. Induction occurs when we observe a fact and ask, "Why is this happening?" In answering this question, we advance one or more tentative explanations (hypotheses). We then use deduction to narrow down the tentative explanations to the most plausible explanation based on logic and reasonable premises (based on our understanding of the phenomenon under study). Researchers must be able to move back and forth between inductive and deductive reasoning if they are to post extensions or modifications to a given model or theory, or built better ones, which are the essence of scientific research.

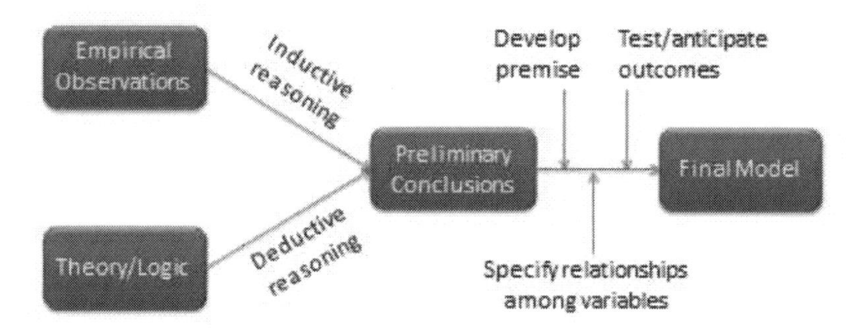

Figure 2.3. The model-building process

Chapter 3

The Research Process

In Chapter 1, we saw that scientific research is the process of acquiring scientific knowledge using the scientific method. But how is such research conducted? This chapter delves into the process of scientific research, and the assumptions and outcomes of the research process.

Paradigms of Social Research

Our design and conduct of research is shaped by our mental models or frames of references that we use to organize our reasoning and observations. These mental models or frames (belief systems) are called **paradigms**. The word "paradigm" was popularized by Thomas Kuhn (1962) in his book *The Structure of Scientific Revolutions*, where he examined the history of the natural sciences to identify patterns of activities that shape the progress of science. Similar ideas are applicable to social sciences as well, where a social reality can be viewed by different people in different ways, which may constrain their thinking and reasoning about the observed phenomenon. For instance, conservatives and liberals tend to have very different perceptions of the role of government in people's lives, and hence, have different opinions on how to solve social problems. Conservatives may believe that lowering taxes is the best way to stimulate a stagnant economy because it increases people's disposable income and spending, which in turn expands business output and employment. In contrast, liberals may believe that governments should invest more directly in job creation programs such as public works and infrastructure projects, which will increase employment and people's ability to consume and drive the economy. Likewise, Western societies place greater emphasis on individual rights, such as one's right to privacy, right of free speech, and right to bear arms. In contrast, Asian societies tend to balance the rights of individuals against the rights of families, organizations, and the government, and therefore tend to be more communal and less individualistic in their policies. Such differences in perspective often lead Westerners to criticize Asian governments for being autocratic, while Asians criticize Western societies for being greedy, having high crime rates, and creating a "cult of the individual." Our personal paradigms are like "colored glasses" that govern how we view the world and how we structure our thoughts about what we see in the world.

Paradigms are often hard to recognize, because they are implicit, assumed, and taken for granted. However, recognizing these paradigms is key to making sense of and reconciling differences in people' perceptions of the same social phenomenon. For instance, why do liberals believe that the best way to improve secondary education is to hire more teachers, but conservatives believe that privatizing education (using such means as school vouchers) are more effective in achieving the same goal? Because conservatives place more faith in competitive markets (i.e., in free competition between schools competing for education dollars), while liberals believe more in labor (i.e., in having more teachers and schools). Likewise, in social science research, if one were to understand why a certain technology was successfully implemented in one organization but failed miserably in another, a researcher looking at the world through a "rational lens" will look for rational explanations of the problem such as inadequate technology or poor fit between technology and the task context where it is being utilized, while another research looking at the same problem through a "social lens" may seek out social deficiencies such as inadequate user training or lack of management support, while those seeing it through a "political lens" will look for instances of organizational politics that may subvert the technology implementation process. Hence, subconscious paradigms often constrain the concepts that researchers attempt to measure, their observations, and their subsequent interpretations of a phenomenon. However, given the complex nature of social phenomenon, it is possible that all of the above paradigms are partially correct, and that a fuller understanding of the problem may require an understanding and application of multiple paradigms.

Two popular paradigms today among social science researchers are positivism and post-positivism. **Positivism**, based on the works of French philosopher Auguste Comte (1798-1857), was the dominant scientific paradigm until the mid-20th century. It holds that science or knowledge creation should be restricted to what can be observed and measured. Positivism tends to rely exclusively on theories that can be directly tested. Though positivism was originally an attempt to separate scientific inquiry from religion (where the precepts could not be objectively observed), positivism led to *empiricism* or a blind faith in observed data and a rejection of any attempt to extend or reason beyond observable facts. Since human thoughts and emotions could not be directly measured, there were not considered to be legitimate topics for scientific research. Frustrations with the strictly empirical nature of positivist philosophy led to the development of **post-positivism** (or postmodernism) during the mid-late 20th century. Post-positivism argues that one can make reasonable inferences about a phenomenon by combining empirical observations with logical reasoning. Post-positivists view science as not certain but probabilistic (i.e., based on many contingencies), and often seek to explore these contingencies to understand social reality better. The post-positivist camp has further fragmented into *subjectivists*, who view the world as a subjective construction of our subjective minds rather than as an objective reality, and *critical realists*, who believe that there is an external reality that is independent of a person's thinking but we can never know such reality with any degree of certainty.

Burrell and Morgan (1979), in their seminal book *Sociological Paradigms and Organizational Analysis*, suggested that the way social science researchers view and study social phenomena is shaped by two fundamental sets of philosophical assumptions: ontology and epistemology. **Ontology** refers to our assumptions about how we see the world, e.g., does the world consist mostly of social order or constant change. **Epistemology** refers to our assumptions about the best way to study the world, e.g., should we use an objective or subjective approach to study social reality. Using these two sets of assumptions, we can categorize social science research as belonging to one of four categories (see Figure 3.1).

If researchers view the world as consisting mostly of social order (ontology) and hence seek to study patterns of ordered events or behaviors, and believe that the best way to study such a world is using objective approach (epistemology) that is independent of the person conducting the observation or interpretation, such as by using standardized data collection tools like surveys, then they are adopting a paradigm of **functionalism**. However, if they believe that the best way to study social order is though the subjective interpretation of participants involved, such as by interviewing different participants and reconciling differences among their responses using their own subjective perspectives, then they are employing an **interpretivism** paradigm. If researchers believe that the world consists of radical change and seek to understand or enact change using an objectivist approach, then they are employing a **radical structuralism** paradigm. If they wish to understand social change using the subjective perspectives of the participants involved, then they are following a **radical humanism** paradigm.

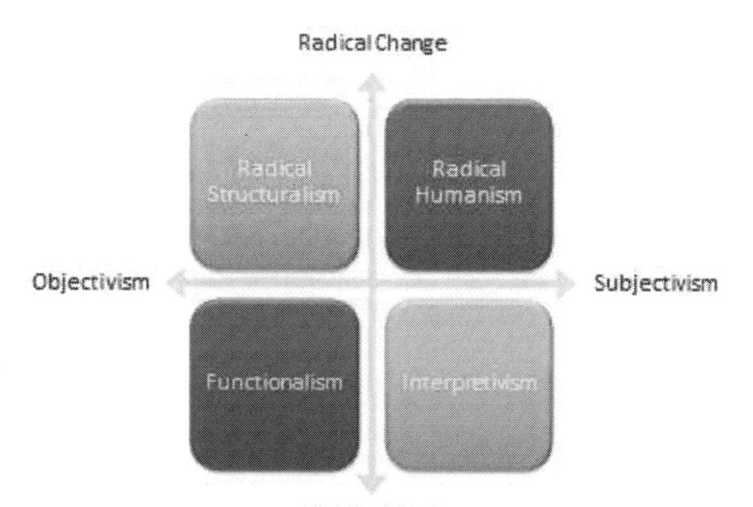

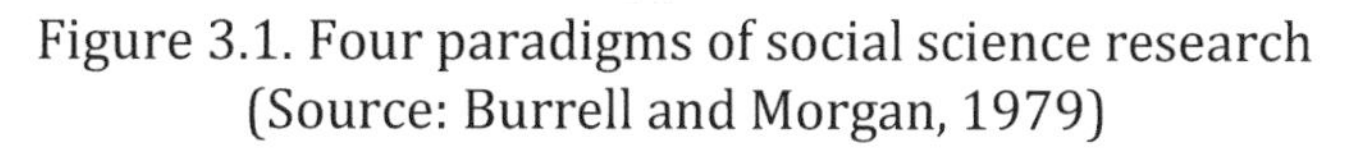

Figure 3.1. Four paradigms of social science research
(Source: Burrell and Morgan, 1979)

To date, the majority of social science research has emulated the natural sciences, and followed the functionalist paradigm. Functionalists believe that social order or patterns can be understood in terms of their functional components, and therefore attempt to break down a problem into small components and studying one or more components in detail using objectivist techniques such as surveys and experimental research. However, with the emergence of post-positivist thinking, a small but growing number of social science researchers are attempting to understand social order using subjectivist techniques such as interviews and ethnographic studies. Radical humanism and radical structuralism continues to represent a negligible proportion of social science research, because scientists are primarily concerned with understanding generalizable patterns of behavior, events, or phenomena, rather than idiosyncratic or changing events. Nevertheless, if you wish to study social change, such as why democratic movements are increasingly emerging in Middle Eastern countries, or why this movement was successful in Tunisia, took a longer path to success in Libya, and is still not successful in Syria, then perhaps radical humanism is the right approach for such a study. Social and organizational phenomena generally consists elements of both order and change. For instance, organizational success depends on formalized business processes, work procedures, and job responsibilities, while being simultaneously constrained by a constantly changing mix of competitors, competing products, suppliers, and customer base in the business environment. Hence, a holistic and more complete understanding of social phenomena such as why are some organizations more successful than others, require an appreciation and application of a multi-paradigmatic approach to research.

Overview of the Research Process

So how do our mental paradigms shape social science research? At its core, all scientific research is an iterative process of observation, rationalization, and validation. In the **observation** phase, we observe a natural or social phenomenon, event, or behavior that interests us. In the **rationalization** phase, we try to make sense of or the observed phenomenon, event, or behavior by logically connecting the different pieces of the puzzle that we observe, which in some cases, may lead to the construction of a theory. Finally, in the **validation** phase, we test our theories using a scientific method through a process of data collection and analysis, and in doing so, possibly modify or extend our initial theory. However, research designs vary based on whether the researcher starts at observation and attempts to rationalize the observations (inductive research), or whether the researcher starts at an ex ante rationalization or a theory and attempts to validate the theory (deductive research). Hence, the observation-rationalization-validation cycle is very similar to the induction-deduction cycle of research discussed in Chapter 1.

Most traditional research tends to be deductive and functionalistic in nature. Figure 3.2 provides a schematic view of such a research project. This figure depicts a series of activities to be performed in functionalist research, categorized into three phases: exploration, research design, and research execution. Note that this generalized design is not a roadmap or flowchart for all research. It applies only to functionalistic research, and it can and should be modified to fit the needs of a specific project.

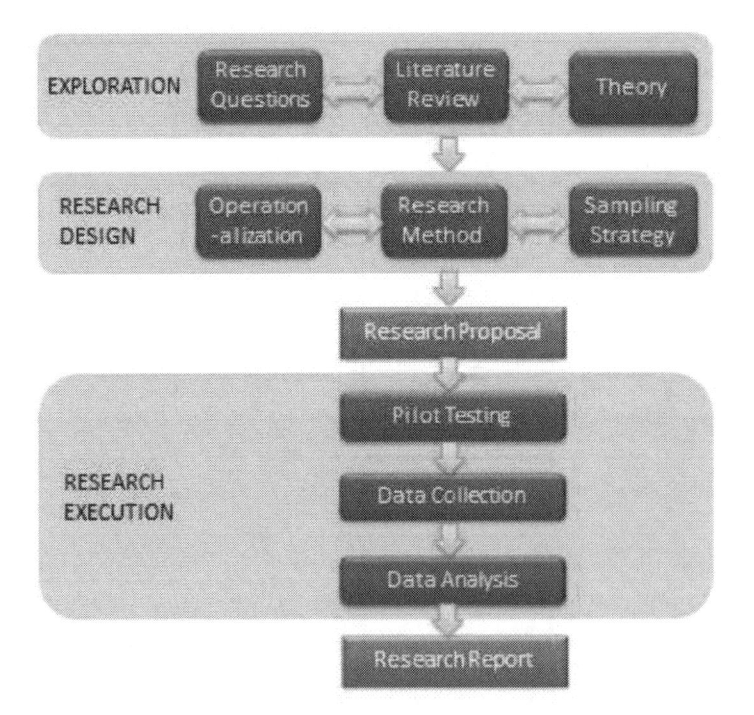

Figure 3.2. Functionalistic research process

The first phase of research is **exploration**. This phase includes exploring and selecting research questions for further investigation, examining the published literature in the area of inquiry to understand the current state of knowledge in that area, and identifying theories that may help answer the research questions of interest.

The first step in the exploration phase is identifying one or more **research questions** dealing with a specific behavior, event, or phenomena of interest. Research questions are specific questions about a behavior, event, or phenomena of interest that you wish to seek answers for in your research. Examples include what factors motivate consumers to purchase goods and services online without knowing the vendors of these goods or services, how can we make high school students more creative, and why do some people commit terrorist acts. Research questions can delve into issues of what, why, how, when, and so forth. More interesting research questions are those that appeal to a broader population (e.g., "how can firms innovate" is a more interesting research question than "how can Chinese firms innovate in the service-sector"), address real and complex problems (in contrast to hypothetical or "toy" problems), and where the answers are not obvious. Narrowly focused research questions (often with a binary yes/no answer) tend to be less useful and less interesting and less suited to capturing the subtle nuances of social phenomena. Uninteresting research questions generally lead to uninteresting and unpublishable research findings.

The next step is to conduct a **literature review** of the domain of interest. The purpose of a literature review is three-fold: (1) to survey the current state of knowledge in the area of inquiry, (2) to identify key authors, articles, theories, and findings in that area, and (3) to identify gaps in knowledge in that research area. Literature review is commonly done today using computerized keyword searches in online databases. Keywords can be combined using "and" and "or" operations to narrow down or expand the search results. Once a shortlist of relevant articles is generated from the keyword search, the researcher must then manually browse through each article, or at least its abstract section, to determine the suitability of that article for a detailed review. Literature reviews should be reasonably complete, and not restricted to a few journals, a few years, or a specific methodology. Reviewed articles may be summarized in the form of tables, and can be further structured using organizing frameworks such as a concept matrix. A well-conducted literature review should indicate whether the initial research questions have already been addressed in the literature (which would obviate the need to study them again), whether there are newer or more interesting research questions available, and whether the original research questions should be modified or changed in light of findings of the literature review. The review can also provide some intuitions or potential answers to the questions of interest and/or help identify theories that have previously been used to address similar questions.

Since functionalist (deductive) research involves theory-testing, the third step is to identify one or more **theories** can help address the desired research questions. While the literature review may uncover a wide range of concepts or constructs potentially related to the phenomenon of interest, a theory will help identify which of these constructs is logically relevant to the target phenomenon and how. Forgoing theories may result in measuring a wide range of less relevant, marginally relevant, or irrelevant constructs, while also minimizing the chances of obtaining results that are meaningful and not by pure chance. In functionalist research, theories can be used as the logical basis for postulating hypotheses for empirical testing. Obviously, not all theories are well-suited for studying all social phenomena. Theories must be carefully selected based on their fit with the target problem and the extent to which their assumptions are consistent with that of the target problem. We will examine theories and the process of theorizing in detail in the next chapter.

The next phase in the research process is **research design**. This process is concerned with creating a blueprint of the activities to take in order to satisfactorily answer the research questions identified in the exploration phase. This includes selecting a research method, operationalizing constructs of interest, and devising an appropriate sampling strategy.

Operationalization is the process of designing precise measures for abstract theoretical constructs. This is a major problem in social science research, given that many of the constructs, such as prejudice, alienation, and liberalism are hard to define, let alone measure accurately. Operationalization starts with specifying an "operational definition" (or "conceptualization") of the constructs of interest. Next, the researcher can search the literature to see if there are existing prevalidated measures matching their operational definition that can be used directly or modified to measure their constructs of interest. If such measures are not available or if existing measures are poor or reflect a different conceptualization than that intended by the researcher, new instruments may have to be designed for measuring those constructs. This means specifying exactly how exactly the desired construct will be measured (e.g., how many items, what items, and so forth). This can easily be a long and laborious process, with multiple rounds of pretests and modifications before the newly designed instrument can be accepted as "scientifically valid." We will discuss operationalization of constructs in a future chapter on measurement.

Simultaneously with operationalization, the researcher must also decide what **research method** they wish to employ for collecting data to address their research questions of interest. Such methods may include quantitative methods such as experiments or survey research or qualitative methods such as case research or action research, or possibly a combination of both. If an experiment is desired, then what is the experimental design? If survey, do you plan a mail survey, telephone survey, web survey, or a combination? For complex, uncertain, and multi-faceted social phenomena, multi-method approaches may be more suitable, which may help leverage the unique strengths of each research method and generate insights that may not be obtained using a single method.

Researchers must also carefully choose the target population from which they wish to collect data, and a **sampling** strategy to select a sample from that population. For instance, should they survey individuals or firms or workgroups within firms? What types of individuals or firms they wish to target? Sampling strategy is closely related to the unit of analysis in a research problem. While selecting a sample, reasonable care should be taken to avoid a biased sample (e.g., sample based on convenience) that may generate biased observations. Sampling is covered in depth in a later chapter.

At this stage, it is often a good idea to write a **research proposal** detailing all of the decisions made in the preceding stages of the research process and the rationale behind each decision. This multi-part proposal should address what research questions you wish to study and why, the prior state of knowledge in this area, theories you wish to employ along with hypotheses to be tested, how to measure constructs, what research method to be employed and why, and desired sampling strategy. Funding agencies typically require such a proposal in order to select the best proposals for funding. Even if funding is not sought for a research project, a proposal may serve as a useful vehicle for seeking feedback from other researchers and identifying potential problems with the research project (e.g., whether some important constructs were missing from the study) before starting data collection. This initial feedback is invaluable because it is often too late to correct critical problems after data is collected in a research study.

Having decided who to study (subjects), what to measure (concepts), and how to collect data (research method), the researcher is now ready to proceed to the **research execution** phase. This includes pilot testing the measurement instruments, data collection, and data analysis.

Pilot testing is an often overlooked but extremely important part of the research process. It helps detect potential problems in your research design and/or instrumentation (e.g., whether the questions asked is intelligible to the targeted sample), and to ensure that the measurement instruments used in the study are reliable and valid measures of the constructs of interest. The pilot sample is usually a small subset of the target population. After a successful pilot testing, the researcher may then proceed with **data collection** using the sampled population. The data collected may be quantitative or qualitative, depending on the research method employed.

Following data collection, the data is analyzed and interpreted for the purpose of drawing conclusions regarding the research questions of interest. Depending on the type of data collected (quantitative or qualitative), **data analysis** may be quantitative (e.g., employ statistical techniques such as regression or structural equation modeling) or qualitative (e.g., coding or content analysis).

The final phase of research involves preparing the final **research report** documenting the entire research process and its findings in the form of a research paper, dissertation, or monograph. This report should outline in detail all the choices made during the research process (e.g., theory used, constructs selected, measures used, research methods, sampling, etc.) and why, as well as the outcomes of each phase of the research process. The research process must be described in sufficient detail so as to allow other researchers to replicate your study, test the findings, or assess whether the inferences derived are scientifically acceptable. Of course, having a ready research proposal will greatly simplify and quicken the process of writing the finished report. Note that research is of no value unless the research process and outcomes are documented for future generations; such documentation is essential for the incremental progress of science.

Common Mistakes in Research

The research process is fraught with problems and pitfalls, and novice researchers often find, after investing substantial amounts of time and effort into a research project, that their research questions were not sufficiently answered, or that the findings were not interesting enough, or that the research was not of "acceptable" scientific quality. Such problems typically result in research papers being rejected by journals. Some of the more frequent mistakes are described below.

Insufficiently motivated research questions. Often times, we choose our "pet" problems that are interesting to us but not to the scientific community at large, i.e., it does not generate new knowledge or insight about the phenomenon being investigated. Because the research process involves a significant investment of time and effort on the researcher's part, the researcher must be certain (and be able to convince others) that the research questions they seek to answer in fact deal with real problems (and not hypothetical problems) that affect a substantial portion of a population and has not been adequately addressed in prior research.

Pursuing research fads. Another common mistake is pursuing "popular" topics with limited shelf life. A typical example is studying technologies or practices that are popular today. Because research takes several years to complete and publish, it is possible that popular interest in these fads may die down by the time the research is completed and submitted for publication. A better strategy may be to study "timeless" topics that have always persisted through the years.

Unresearchable problems. Some research problems may not be answered adequately based on observed evidence alone, or using currently accepted methods and procedures. Such problems are best avoided. However, some unresearchable, ambiguously defined problems may be modified or fine tuned into well-defined and useful researchable problems.

Favored research methods. Many researchers have a tendency to recast a research problem so that it is amenable to their favorite research method (e.g., survey research). This is an unfortunate trend. Research methods should be chosen to best fit a research problem, and not the other way around.

Blind data mining. Some researchers have the tendency to collect data first (using instruments that are already available), and then figure out what to do with it. Note that data collection is only one step in a long and elaborate process of planning, designing, and executing research. In fact, a series of other activities are needed in a research process prior to data collection. If researchers jump into data collection without such elaborate planning, the data collected will likely be irrelevant, imperfect, or useless, and their data collection efforts may be entirely wasted. An abundance of data cannot make up for deficits in research planning and design, and particularly, for the lack of interesting research questions.

Chapter 4

Theories in Scientific Research

As we know from previous chapters, science is knowledge represented as a collection of "theories" derived using the scientific method. In this chapter, we will examine what is a theory, why do we need theories in research, what are the building blocks of a theory, how to evaluate theories, how can we apply theories in research, and also presents illustrative examples of five theories frequently used in social science research.

Theories

Theories are explanations of a natural or social behavior, event, or phenomenon. More formally, a scientific theory is a system of constructs (concepts) and propositions (relationships between those constructs) that collectively presents a logical, systematic, and coherent explanation of a phenomenon of interest within some assumptions and boundary conditions (Bacharach 1989).[1]

Theories should explain why things happen, rather than just describe or predict. Note that it is possible to predict events or behaviors using a set of predictors, without necessarily explaining why such events are taking place. For instance, market analysts predict fluctuations in the stock market based on market announcements, earnings reports of major companies, and new data from the Federal Reserve and other agencies, based on previously observed *correlations*. Prediction requires only correlations. In contrast, explanations require *causations*, or understanding of cause-effect relationships. Establishing causation requires three conditions: (1) correlations between two constructs, (2) temporal precedence (the cause must precede the effect in time), and (3) rejection of alternative hypotheses (through testing). Scientific theories are different from theological, philosophical, or other explanations in that scientific theories can be empirically tested using scientific methods.

Explanations can be idiographic or nomothetic. **Idiographic explanations** are those that explain a single situation or event in idiosyncratic detail. For example, you did poorly on an exam because: (1) you forgot that you had an exam on that day, (2) you arrived late to the exam due to a traffic jam, (3) you panicked midway through the exam, (4) you had to work late the previous evening and could not study for the exam, or even (5) your dog ate your text book. The explanations may be detailed, accurate, and valid, but they may not apply to other similar situations, even involving the same person, and are hence not generalizable. In contrast,

[1] Bacharach, S. B. (1989). "Organizational Theories: Some Criteria for Evaluation," *Academy of Management Review* (14:4), 496-515.

nomothetic explanations seek to explain a class of situations or events rather than a specific situation or event. For example, students who do poorly in exams do so because they did not spend adequate time preparing for exams or that they suffer from nervousness, attention-deficit, or some other medical disorder. Because nomothetic explanations are designed to be generalizable across situations, events, or people, they tend to be less precise, less complete, and less detailed. However, they explain economically, using only a few explanatory variables. Because theories are also intended to serve as generalized explanations for patterns of events, behaviors, or phenomena, theoretical explanations are generally nomothetic in nature.

While understanding theories, it is also important to understand what theory is not. Theory is not data, facts, typologies, taxonomies, or empirical findings. A collection of facts is not a theory, just as a pile of stones is not a house. Likewise, a collection of constructs (e.g., a typology of constructs) is not a theory, because theories must go well beyond constructs to include propositions, explanations, and boundary conditions. Data, facts, and findings operate at the empirical or observational level, while theories operate at a conceptual level and are based on logic rather than observations.

There are many benefits to using theories in research. First, theories provide the underlying logic of the occurrence of natural or social phenomenon by explaining what are the key drivers and key outcomes of the target phenomenon and why, and what underlying processes are responsible driving that phenomenon. Second, they aid in sense-making by helping us synthesize prior empirical findings within a theoretical framework and reconcile contradictory findings by discovering contingent factors influencing the relationship between two constructs in different studies. Third, theories provide guidance for future research by helping identify constructs and relationships that are worthy of further research. Fourth, theories can contribute to cumulative knowledge building by bridging gaps between other theories and by causing existing theories to be reevaluated in a new light.

However, theories can also have their own share of limitations. As simplified explanations of reality, theories may not always provide adequate explanations of the phenomenon of interest based on a limited set of constructs and relationships. Theories are designed to be simple and parsimonious explanations, while reality may be significantly more complex. Furthermore, theories may impose blinders or limit researchers' "range of vision," causing them to miss out on important concepts that are not defined by the theory.

Building Blocks of a Theory

David Whetten (1989) suggests that there are four building blocks of a theory: constructs, propositions, logic, and boundary conditions/assumptions. Constructs capture the "what" of theories (i.e., what concepts are important for explaining a phenomenon), propositions capture the "how" (i.e., how are these concepts related to each other), logic represents the "why" (i.e., why are these concepts related), and boundary conditions/assumptions examines the "who, when, and where" (i.e., under what circumstances will these concepts and relationships work). Though constructs and propositions were previously discussed in Chapter 2, we describe them again here for the sake of completeness.

Constructs are abstract concepts specified at a high level of abstraction that are chosen specifically to explain the phenomenon of interest. Recall from Chapter 2 that constructs may be unidimensional (i.e., embody a single concept), such as weight or age, or multi-dimensional (i.e., embody multiple underlying concepts), such as personality or culture. While some constructs, such as age, education, and firm size, are easy to understand, others, such as creativity, prejudice, and organizational agility, may be more complex and abstruse, and still others such as trust, attitude, and learning, may represent temporal tendencies rather than steady states. Nevertheless, all constructs must have clear and unambiguous operational definition that should specify exactly how the construct will be measured and at what level of analysis (individual, group, organizational, etc.). Measurable representations of abstract constructs are called **variables**. For instance, intelligence quotient (IQ score) is a variable that is purported to measure an abstract construct called intelligence. As noted earlier, scientific research proceeds along two planes: a theoretical plane and an empirical plane. Constructs are conceptualized at the theoretical plane, while variables are operationalized and measured at the empirical (observational) plane. Furthermore, variables may be independent, dependent, mediating, or moderating, as discussed in Chapter 2. The distinction between constructs (conceptualized at the theoretical level) and variables (measured at the empirical level) is shown in Figure 4.1.

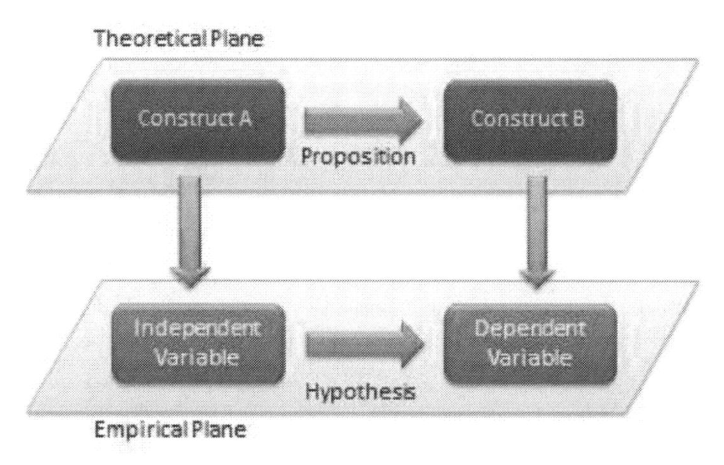

Figure 4.1. Distinction between theoretical and empirical concepts

Propositions are associations postulated between constructs based on deductive logic. Propositions are stated in declarative form and should ideally indicate a cause-effect relationship (e.g., if X occurs, then Y will follow). Note that propositions may be conjectural but MUST be testable, and should be rejected if they are not supported by empirical observations. However, like constructs, propositions are stated at the theoretical level, and they can only be tested by examining the corresponding relationship between measurable variables of those constructs. The empirical formulation of propositions, stated as relationships between variables, is called **hypotheses**. The distinction between propositions (formulated at the theoretical level) and hypotheses (tested at the empirical level) is depicted in Figure 4.1.

The third building block of a theory is the **logic** that provides the basis for justifying the propositions as postulated. Logic acts like a "glue" that connects the theoretical constructs and provides meaning and relevance to the relationships between these constructs. Logic also represents the "explanation" that lies at the core of a theory. Without logic, propositions will be ad hoc, arbitrary, and meaningless, and cannot be tied into a cohesive "system of propositions" that is the heart of any theory.

Finally, all theories are constrained by **assumptions** about values, time, and space, and **boundary conditions** that govern where the theory can be applied and where it cannot be applied. For example, many economic theories assume that human beings are rational (or boundedly rational) and employ utility maximization based on cost and benefit expectations as a way of understand human behavior. In contrast, political science theories assume that people are more political than rational, and try to position themselves in their professional or personal environment in a way that maximizes their power and control over others. Given the nature of their underlying assumptions, economic and political theories are not directly comparable, and researchers should not use economic theories if their objective is to understand the power structure or its evolution in a organization. Likewise, theories may have implicit cultural assumptions (e.g., whether they apply to individualistic or collective cultures), temporal assumptions (e.g., whether they apply to early stages or later stages of human behavior), and spatial assumptions (e.g., whether they apply to certain localities but not to others). If a theory is to be properly used or tested, all of its implicit assumptions that form the boundaries of that theory must be properly understood. Unfortunately, theorists rarely state their implicit assumptions clearly, which leads to frequent misapplications of theories to problem situations in research.

Attributes of a Good Theory

Theories are simplified and often partial explanations of complex social reality. As such, there can be good explanations or poor explanations, and consequently, there can be good theories or poor theories. How can we evaluate the "goodness" of a given theory? Different criteria have been proposed by different researchers, the more important of which are listed below:

- **Logical consistency:** Are the theoretical constructs, propositions, boundary conditions, and assumptions logically consistent with each other? If some of these "building blocks" of a theory are inconsistent with each other (e.g., a theory assumes rationality, but some constructs represent non-rational concepts), then the theory is a poor theory.

- **Explanatory power:** How much does a given theory explain (or predict) reality? Good theories obviously explain the target phenomenon better than rival theories, as often measured by variance explained (R-square) value in regression equations.

- **Falsifiability:** British philosopher Karl Popper stated in the 1940's that for theories to be valid, they must be falsifiable. Falsifiability ensures that the theory is potentially disprovable, if empirical data does not match with theoretical propositions, which allows for their empirical testing by researchers. In other words, theories cannot be theories unless they can be empirically testable. Tautological statements, such as "a day with high temperatures is a hot day" are not empirically testable because a hot day is defined (and measured) as a day with high temperatures, and hence, such statements cannot be viewed as a theoretical proposition. Falsifiability requires presence of rival explanations it ensures that the constructs are adequately measurable, and so forth. However, note that saying that a theory is falsifiable is not the same as saying that a theory should be falsified. If a theory is indeed falsified based on empirical evidence, then it was probably a poor theory to begin with!

- **Parsimony:** Parsimony examines how much of a phenomenon is explained with how few variables. The concept is attributed to 14th century English logician Father William of Ockham (and hence called "Ockham's razor" or "Occam's razor), which states that among competing explanations that sufficiently explain the observed evidence, the simplest theory (i.e., one that uses the smallest number of variables or makes the fewest assumptions) is the best. Explanation of a complex social phenomenon can always be increased by adding more and more constructs. However, such approach defeats the purpose of having a theory, which are intended to be "simplified" and generalizable explanations of reality. Parsimony relates to the degrees of freedom in a given theory. Parsimonious theories have higher degrees of freedom, which allow them to be more easily generalized to other contexts, settings, and populations.

Approaches to Theorizing

How do researchers build theories? Steinfeld and Fulk (1990)[2] recommend four such approaches. The first approach is to build theories inductively based on observed patterns of events or behaviors. Such approach is often called "grounded theory building", because the theory is grounded in empirical observations. This technique is heavily dependent on the observational and interpretive abilities of the researcher, and the resulting theory may be subjective and non-confirmable. Furthermore, observing certain patterns of events will not necessarily make a theory, unless the researcher is able to provide consistent explanations for the observed patterns. We will discuss the grounded theory approach in a later chapter on qualitative research.

The second approach to theory building is to conduct a bottom-up conceptual analysis to identify different sets of predictors relevant to the phenomenon of interest using a predefined framework. One such framework may be a simple input-process-output framework, where the researcher may look for different categories of inputs, such as individual, organizational, and/or technological factors potentially related to the phenomenon of interest (the output), and describe the underlying processes that link these factors to the target phenomenon. This is also an inductive approach that relies heavily on the inductive abilities of the researcher, and interpretation may be biased by researcher's prior knowledge of the phenomenon being studied.

The third approach to theorizing is to extend or modify existing theories to explain a new context, such as by extending theories of individual learning to explain organizational learning. While making such an extension, certain concepts, propositions, and/or boundary conditions of the old theory may be retained and others modified to fit the new context. This deductive approach leverages the rich inventory of social science theories developed by prior theoreticians, and is an efficient way of building new theories by building on existing ones.

The fourth approach is to apply existing theories in entirely new contexts by drawing upon the structural similarities between the two contexts. This approach relies on reasoning by analogy, and is probably the most creative way of theorizing using a deductive approach. For instance, Markus (1987)[3] used analogic similarities between a nuclear explosion and uncontrolled growth of networks or network-based businesses to propose a critical mass theory of network growth. Just as a nuclear explosion requires a critical mass of radioactive material to sustain a nuclear explosion, Markus suggested that a network requires a critical mass of users to sustain its growth, and without such critical mass, users may leave the network, causing an eventual demise of the network.

---

[2] Steinfield, C.W. and Fulk, J. (1990). "The Theory Imperative," in *Organizations and Communications Technology*, J. Fulk and C. W. Steinfield (eds.), Newbury Park, CA: Sage Publications.

[3] Markus, M. L. (1987). "Toward a 'Critical Mass' Theory of Interactive Media: Universal Access, Interdependence, and Diffusion," *Communication Research* (14:5), 491-511.

Examples of Social Science Theories

In this section, we present brief overviews of a few illustrative theories from different social science disciplines. These theories explain different types of social behaviors, using a set of constructs, propositions, boundary conditions, assumptions, and underlying logic. Note that the following represents just a simplistic introduction to these theories; readers are advised to consult the original sources of these theories for more details and insights on each theory.

Agency Theory. Agency theory (also called principal-agent theory), a classic theory in the organizational economics literature, was originally proposed by Ross (1973)[4] to explain two-party relationships (such as those between an employer and its employees, between organizational executives and shareholders, and between buyers and sellers) whose goals are not congruent with each other. The goal of agency theory is to specify optimal contracts and the conditions under which such contracts may help minimize the effect of goal incongruence. The core assumptions of this theory are that human beings are self-interested individuals, boundedly rational, and risk-averse, and the theory can be applied at the individual or organizational level.

The two parties in this theory are the principal and the agent; the principal employs the agent to perform certain tasks on its behalf. While the principal's goal is quick and effective completion of the assigned task, the agent's goal may be working at its own pace, avoiding risks, and seeking self-interest (such as personal pay) over corporate interests. Hence, the goal incongruence. Compounding the nature of the problem may be information asymmetry problems caused by the principal's inability to adequately observe the agent's behavior or accurately evaluate the agent's skill sets. Such asymmetry may lead to agency problems where the agent may not put forth the effort needed to get the task done (the *moral hazard* problem) or may misrepresent its expertise or skills to get the job but not perform as expected (the *adverse selection* problem). Typical contracts that are behavior-based, such as a monthly salary, cannot overcome these problems. Hence, agency theory recommends using outcome-based contracts, such as a commissions or a fee payable upon task completion, or mixed contracts that combine behavior-based and outcome-based incentives. An employee stock option plans are is an example of an outcome-based contract while employee pay is a behavior-based contract. Agency theory also recommends tools that principals may employ to improve the efficacy of behavior-based contracts, such as investing in monitoring mechanisms (such as hiring supervisors) to counter the information asymmetry caused by moral hazard, designing renewable contracts contingent on agent's performance (performance assessment makes the contract partially outcome-based), or by improving the structure of the assigned task to make it more programmable and therefore more observable.

Theory of Planned Behavior. Postulated by Azjen (1991)[5], the theory of planned behavior (TPB) is a generalized theory of human behavior in the social psychology literature that can be used to study a wide range of individual behaviors. It presumes that individual behavior represents conscious reasoned choice, and is shaped by cognitive thinking and social pressures. The theory postulates that behaviors are based on one's intention regarding that behavior, which in turn is a function of the person's attitude toward the behavior, subjective

---

[4] Ross, S. A. (1973). "The Economic Theory of Agency: The Principal's Problem," *American Economic Review* (63:2), 134-139.

[5] Ajzen, I. (1991). "The Theory of Planned Behavior," *Organizational Behavior and Human Decision Processes* (50), 179-211.

norm regarding that behavior, and perception of control over that behavior (see Figure 4.2). Attitude is defined as the individual's overall positive or negative feelings about performing the behavior in question, which may be assessed as a summation of one's beliefs regarding the different consequences of that behavior, weighted by the desirability of those consequences. Subjective norm refers to one's perception of whether people important to that person expect the person to perform the intended behavior, and represented as a weighted combination of the expected norms of different referent groups such as friends, colleagues, or supervisors at work. Behavioral control is one's perception of internal or external controls constraining the behavior in question. Internal controls may include the person's ability to perform the intended behavior (self-efficacy), while external control refers to the availability of external resources needed to perform that behavior (facilitating conditions). TPB also suggests that sometimes people may intend to perform a given behavior but lack the resources needed to do so, and therefore suggests that posits that behavioral control can have a direct effect on behavior, in addition to the indirect effect mediated by intention.

TPB is an extension of an earlier theory called the theory of reasoned action, which included attitude and subjective norm as key drivers of intention, but not behavioral control. The latter construct was added by Ajzen in TPB to account for circumstances when people may have incomplete control over their own behaviors (such as not having high-speed Internet access for web surfing).

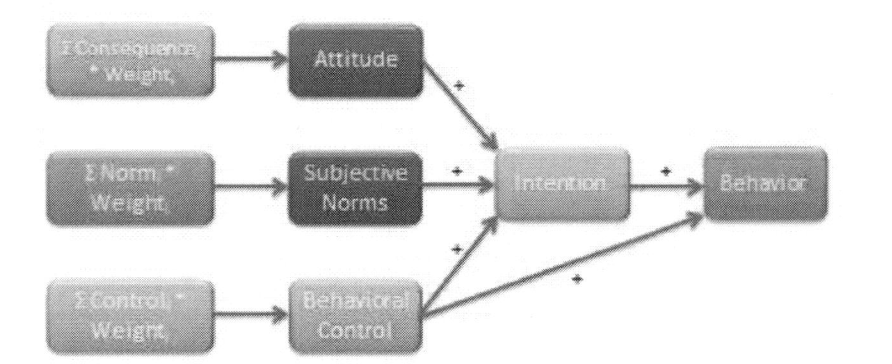

Figure 4.2. Theory of planned behavior

Innovation diffusion theory. Innovation diffusion theory (IDT) is a seminal theory in the communications literature that explains how innovations are adopted within a population of potential adopters. The concept was first studied by French sociologist Gabriel Tarde, but the theory was developed by Everett Rogers in 1962 based on observations of 508 diffusion studies. The four key elements in this theory are: innovation, communication channels, time, and social system. Innovations may include new technologies, new practices, or new ideas, and adopters may be individuals or organizations. At the macro (population) level, IDT views innovation diffusion as a process of communication where people in a social system learn about a new innovation and its potential benefits through communication channels (such as mass media or prior adopters) and are persuaded to adopt it. Diffusion is a temporal process; the diffusion process starts off slow among a few early adopters, then picks up speed as the innovation is adopted by the mainstream population, and finally slows down as the adopter population reaches saturation. The cumulative adoption pattern therefore an S-shaped curve, as shown in Figure 4.3, and the adopter distribution represents a normal distribution. All adopters are not identical, and adopters can be classified into innovators, early adopters, early majority, late majority, and laggards based on their time of their adoption. The rate of diffusion also depends on characteristics of the social system such as the presence of opinion leaders (experts whose opinions are valued by others) and change agents (people who influence others' behaviors).

At the micro (adopter) level, Rogers (1995)[6] suggests that innovation adoption is a process consisting of five stages: (1) knowledge: when adopters first learn about an innovation from mass-media or interpersonal channels, (2) persuasion: when they are persuaded by prior adopters to try the innovation, (3) decision: their decision to accept or reject the innovation, (4) implementation: their initial utilization of the innovation, and (5) confirmation: their decision to continue using it to its fullest potential (see Figure 4.4). Five innovation characteristics are presumed to shape adopters' innovation adoption decisions: (1) relative advantage: the expected benefits of an innovation relative to prior innovations, (2) compatibility: the extent to which the innovation fits with the adopter's work habits, beliefs, and values, (3) complexity: the extent to which the innovation is difficult to learn and use, (4) trialability: the extent to which the innovation can be tested on a trial basis, and (5) observability: the extent to which the results of using the innovation can be clearly observed. The last two characteristics have since been dropped from many innovation studies. Complexity is negatively correlated to innovation adoption, while the other four factors are positively correlated. Innovation adoption also depends on personal factors such as the adopter's risk-taking propensity, education level, cosmopolitanism, and communication influence. Early adopters are venturesome, well educated, and rely more on mass media for information about the innovation, while later adopters rely more on interpersonal sources (such as friends and family) as their primary source of information. IDT has been criticized for having a "pro-innovation bias," that is for presuming that all innovations are beneficial and will be eventually diffused across the entire population, and because it does not allow for inefficient innovations such as fads or fashions to die off quickly without being adopted by the entire population or being replaced by better innovations.

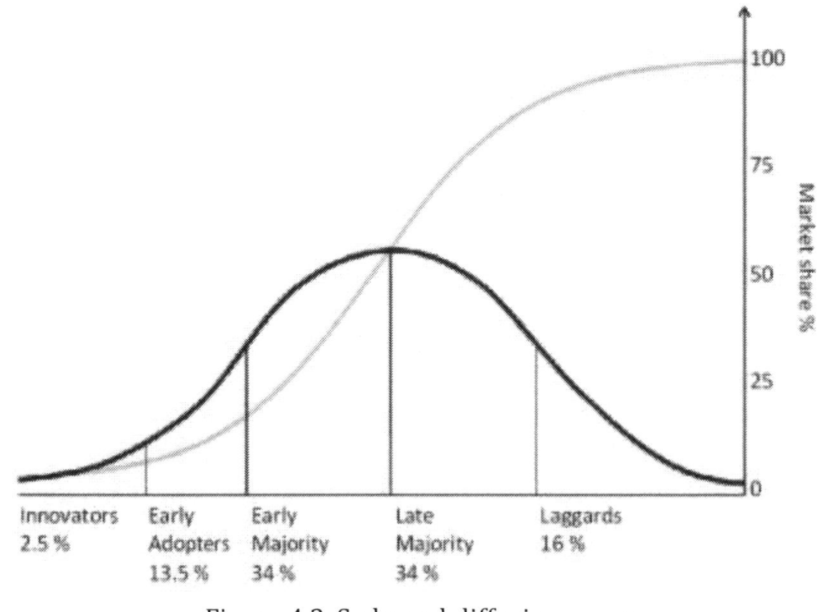

Figure 4.3. S-shaped diffusion curve

---

[6] Rogers, E. (1962). *Diffusion of Innovations*. New York: The Free Press. Other editions 1983, 1996, 2005.

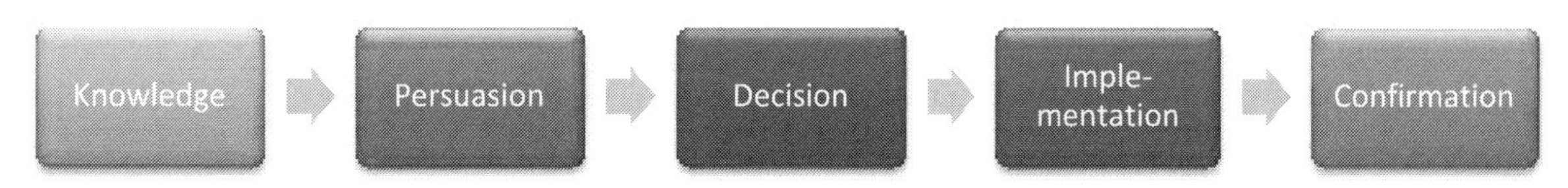

Figure 4.4. Innovation adoption process

Elaboration Likelihood Model. Developed by Petty and Cacioppo (1986)[7], the elaboration likelihood model (ELM) is a dual-process theory of attitude formation or change in the psychology literature. It explains how individuals can be influenced to change their attitude toward a certain object, events, or behavior and the relative efficacy of such change strategies. The ELM posits that one's attitude may be shaped by two "routes" of influence, the central route and the peripheral route, which differ in the amount of thoughtful information processing or "elaboration" required of people (see Figure 4.5). The central route requires a person to think about issue-related arguments in an informational message and carefully scrutinize the merits and relevance of those arguments, before forming an informed judgment about the target object. In the peripheral route, subjects rely on external "cues" such as number of prior users, endorsements from experts, or likeability of the endorser, rather than on the quality of arguments, in framing their attitude towards the target object. The latter route is less cognitively demanding, and the routes of attitude change are typically operationalized in the ELM using the *argument quality* and *peripheral cues* constructs respectively.

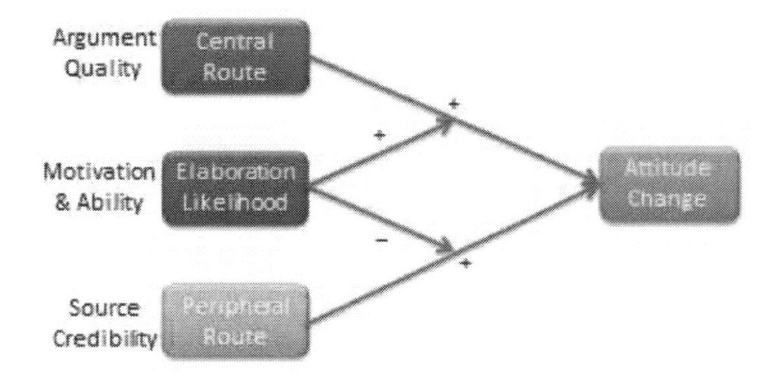

Figure 4.5. Elaboration likelihood model

Whether people will be influenced by the central or peripheral routes depends upon their ability and motivation to elaborate the central merits of an argument. This ability and motivation to elaborate is called *elaboration likelihood*. People in a state of high elaboration likelihood (high ability and high motivation) are more likely to thoughtfully process the information presented and are therefore more influenced by argument quality, while those in the low elaboration likelihood state are more motivated by peripheral cues. Elaboration likelihood is a situational characteristic and not a personal trait. For instance, a doctor may employ the central route for diagnosing and treating a medical ailment (by virtue of his or her expertise of the subject), but may rely on peripheral cues from auto mechanics to understand

[7] Petty, R. E., and Cacioppo, J. T. (1986). *Communication and Persuasion: Central and Peripheral Routes to Attitude Change*. New York: Springer-Verlag.

the problems with his car. As such, the theory has widespread implications about how to enact attitude change toward new products or ideas and even social change.

General Deterrence Theory. Two utilitarian philosophers of the eighteenth century, Cesare Beccaria and Jeremy Bentham, formulated General Deterrence Theory (GDT) as both an explanation of crime and a method for reducing it. GDT examines why certain individuals engage in deviant, anti-social, or criminal behaviors. This theory holds that people are fundamentally rational (for both conforming and deviant behaviors), and that they freely choose deviant behaviors based on a rational cost-benefit calculation. Because people naturally choose utility-maximizing behaviors, deviant choices that engender personal gain or pleasure can be controlled by increasing the costs of such behaviors in the form of punishments (countermeasures) as well as increasing the probability of apprehension. Swiftness, severity, and certainty of punishments are the key constructs in GDT.

While classical positivist research in criminology seeks generalized causes of criminal behaviors, such as poverty, lack of education, psychological conditions, and recommends strategies to rehabilitate criminals, such as by providing them job training and medical treatment, GDT focuses on the criminal decision making process and situational factors that influence that process. Hence, a criminal's personal situation (such as his personal values, his affluence, and his need for money) and the environmental context (such as how protected is the target, how efficient is the local police, how likely are criminals to be apprehended) play key roles in this decision making process. The focus of GDT is not how to rehabilitate criminals and avert future criminal behaviors, but how to make criminal activities less attractive and therefore prevent crimes. To that end, "target hardening" such as installing deadbolts and building self-defense skills, legal deterrents such as eliminating parole for certain crimes, "three strikes law" (mandatory incarceration for three offenses, even if the offenses are minor and not worth imprisonment), and the death penalty, increasing the chances of apprehension using means such as neighborhood watch programs, special task forces on drugs or gang-related crimes, and increased police patrols, and educational programs such as highly visible notices such as "Trespassers will be prosecuted" are effective in preventing crimes. This theory has interesting implications not only for traditional crimes, but also for contemporary white-collar crimes such as insider trading, software piracy, and illegal sharing of music.

Chapter 5

Research Design

Research design is a comprehensive plan for data collection in an empirical research project. It is a "blueprint" for empirical research aimed at answering specific research questions or testing specific hypotheses, and must specify at least three processes: (1) the data collection process, (2) the instrument development process, and (3) the sampling process. The instrument development and sampling processes are described in next two chapters, and the data collection process (which is often loosely called "research design") is introduced in this chapter and is described in further detail in Chapters 9-12.

Broadly speaking, data collection methods can be broadly grouped into two categories: positivist and interpretive. **Positivist methods**, such as laboratory experiments and survey research, are aimed at theory (or hypotheses) testing, while interpretive methods, such as action research and ethnography, are aimed at theory building. Positivist methods employ a deductive approach to research, starting with a theory and testing theoretical postulates using empirical data. In contrast, **interpretive methods** employ an inductive approach that starts with data and tries to derive a theory about the phenomenon of interest from the observed data. Often times, these methods are incorrectly equated with quantitative and qualitative research. Quantitative and qualitative methods refers to the type of data being collected (quantitative data involve numeric scores, metrics, and so on, while qualitative data includes interviews, observations, and so forth) and analyzed (i.e., using quantitative techniques such as regression or qualitative techniques such as coding). Positivist research uses predominantly quantitative data, but can also use qualitative data. Interpretive research relies heavily on qualitative data, but can sometimes benefit from including quantitative data as well. Sometimes, joint use of qualitative and quantitative data may help generate unique insight into a complex social phenomenon that are not available from either types of data alone, and hence, **mixed-mode designs** that combine qualitative and quantitative data are often highly desirable.

Key Attributes of a Research Design

The quality of research designs can be defined in terms of four key design attributes: internal validity, external validity, construct validity, and statistical conclusion validity.

Internal validity, also called causality, examines whether the observed change in a dependent variable is indeed caused by a corresponding change in hypothesized independent variable, and not by variables extraneous to the research context. Causality requires three conditions: (1) covariation of cause and effect (i.e., if cause happens, then effect also happens; and if cause does not happen, effect does not happen), (2) temporal precedence: cause must precede effect in time, (3) no plausible alternative explanation (or spurious correlation). Certain research designs, such as laboratory experiments, are strong in internal validity by virtue of their ability to manipulate the independent variable (cause) via a treatment and observe the effect (dependent variable) of that treatment after a certain point in time, while controlling for the effects of extraneous variables. Other designs, such as field surveys, are poor in internal validity because of their inability to manipulate the independent variable (cause), and because cause and effect are measured at the same point in time which defeats temporal precedence making it equally likely that the expected effect might have influenced the expected cause rather than the reverse. Although higher in internal validity compared to other methods, laboratory experiments are, by no means, immune to threats of internal validity, and are susceptible to history, testing, instrumentation, regression, and other threats that are discussed later in the chapter on experimental designs. Nonetheless, different research designs vary considerably in their respective level of internal validity.

External validity or generalizability refers to whether the observed associations can be generalized from the sample to the population (population validity), or to other people, organizations, contexts, or time (ecological validity). For instance, can results drawn from a sample of financial firms in the United States be generalized to the population of financial firms (population validity) or to other firms within the United States (ecological validity)? Survey research, where data is sourced from a wide variety of individuals, firms, or other units of analysis, tends to have broader generalizability than laboratory experiments where artificially contrived treatments and strong control over extraneous variables render the findings less generalizable to real-life settings where treatments and extraneous variables cannot be controlled. The variation in internal and external validity for a wide range of research designs are shown in Figure 5.1.

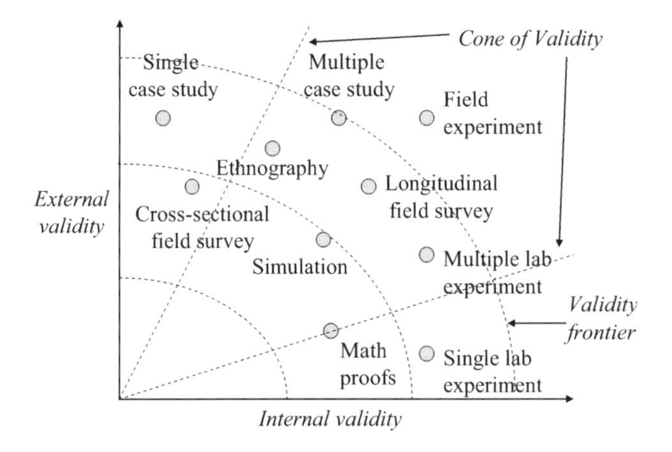

Figure 5.1. Internal and external validity

Some researchers claim that there is a tradeoff between internal and external validity: higher external validity can come only at the cost of internal validity and vice-versa. But this is not always the case. Research designs such as field experiments, longitudinal field surveys, and multiple case studies have higher degrees of both internal and external validities. Personally, I prefer research designs that have reasonable degrees of both internal and external validities, i.e., those that fall within the cone of validity shown in Figure 5.1. But this should not suggest that designs outside this cone are any less useful or valuable. Researchers' choice of designs is ultimately a matter of their personal preference and competence, and the level of internal and external validity they desire.

Construct validity examines how well a given measurement scale is measuring the theoretical construct that it is expected to measure. Many constructs used in social science research such as empathy, resistance to change, and organizational learning are difficult to define, much less measure. For instance, construct validity must assure that a measure of empathy is indeed measuring empathy and not compassion, which may be difficult since these constructs are somewhat similar in meaning. Construct validity is assessed in positivist research based on correlational or factor analysis of pilot test data, as described in the next chapter.

Statistical conclusion validity examines the extent to which conclusions derived using a statistical procedure is valid. For example, it examines whether the right statistical method was used for hypotheses testing, whether the variables used meet the assumptions of that statistical test (such as sample size or distributional requirements), and so forth. Because interpretive research designs do not employ statistical test, statistical conclusion validity is not applicable for such analysis. The different kinds of validity and where they exist at the theoretical/empirical levels are illustrated in Figure 5.2.

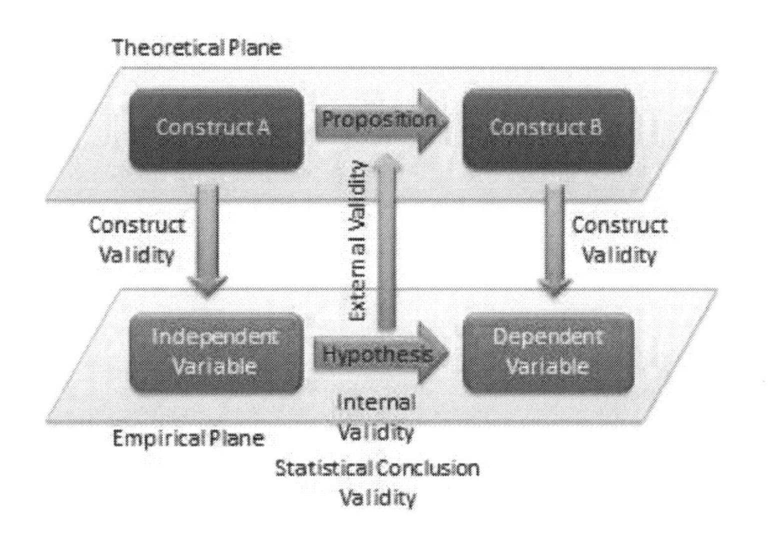

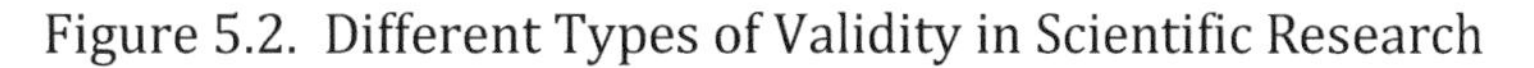
Figure 5.2. Different Types of Validity in Scientific Research

Improving Internal and External Validity

The best research designs are those that can assure high levels of internal and external validity. Such designs would guard against spurious correlations, inspire greater faith in the hypotheses testing, and ensure that the results drawn from a small sample are generalizable to the population at large. Controls are required to assure internal validity (causality) of research designs, and can be accomplished in four ways: (1) manipulation, (2) elimination, (3) inclusion, and (4) statistical control, and (5) randomization.

In **manipulation**, the researcher manipulates the independent variables in one or more levels (called "treatments"), and compares the effects of the treatments against a control group where subjects do not receive the treatment. Treatments may include a new drug or different dosage of drug (for treating a medical condition), a, a teaching style (for students), and so forth. This type of control is achieved in experimental or quasi-experimental designs but not in non-experimental designs such as surveys. Note that if subjects cannot distinguish adequately between different levels of treatment manipulations, their responses across treatments may not be different, and manipulation would fail.

The **elimination** technique relies on eliminating extraneous variables by holding them constant across treatments, such as by restricting the study to a single gender or a single socio-economic status. In the **inclusion** technique, the role of extraneous variables is considered by including them in the research design and separately estimating their effects on the dependent variable, such as via factorial designs where one factor is gender (male versus female). Such technique allows for greater generalizability but also requires substantially larger samples. In **statistical control**, extraneous variables are measured and used as covariates during the statistical testing process.

Finally, the **randomization** technique is aimed at canceling out the effects of extraneous variables through a process of random sampling, if it can be assured that these effects are of a random (non-systematic) nature. Two types of randomization are: (1) **random selection**, where a sample is selected randomly from a population, and (2) **random assignment**, where subjects selected in a non-random manner are randomly assigned to treatment groups.

Randomization also assures external validity, allowing inferences drawn from the sample to be generalized to the population from which the sample is drawn. Note that random assignment is mandatory when random selection is not possible because of resource or access constraints. However, generalizability across populations is harder to ascertain since populations may differ on multiple dimensions and you can only control for few of those dimensions.

Popular Research Designs

As noted earlier, research designs can be classified into two categories – positivist and interpretive – depending how their goal in scientific research. Positivist designs are meant for theory testing, while interpretive designs are meant for theory building. Positivist designs seek generalized patterns based on an objective view of reality, while interpretive designs seek subjective interpretations of social phenomena from the perspectives of the subjects involved. Some popular examples of positivist designs include laboratory experiments, field experiments, field surveys, secondary data analysis, and case research while examples of interpretive designs include case research, phenomenology, and ethnography. Note that case research can be used for theory building or theory testing, though not at the same time. Not all techniques are suited for all kinds of scientific research. Some techniques such as focus groups are best suited for exploratory research, others such as ethnography are best for descriptive research, and still others such as laboratory experiments are ideal for explanatory research. Following are brief descriptions of some of these designs. Additional details are provided in Chapters 9-12.

Experimental studies are those that are intended to test cause-effect relationships (hypotheses) in a tightly controlled setting by separating the cause from the effect in time, administering the cause to one group of subjects (the "treatment group") but not to another group ("control group"), and observing how the mean effects vary between subjects in these two groups. For instance, if we design a laboratory experiment to test the efficacy of a new drug in treating a certain ailment, we can get a random sample of people afflicted with that ailment, randomly assign them to one of two groups (treatment and control groups), administer the drug to subjects in the treatment group, but only give a placebo (e.g., a sugar pill with no medicinal value). More complex designs may include multiple treatment groups, such as low versus high dosage of the drug, multiple treatments, such as combining drug administration with dietary interventions. In a **true experimental design**, subjects must be randomly assigned between each group. If random assignment is not followed, then the design becomes **quasi-experimental**. Experiments can be conducted in an artificial or laboratory setting such as at a university (laboratory experiments) or in field settings such as in an organization where the phenomenon of interest is actually occurring (field experiments). Laboratory experiments allow the researcher to isolate the variables of interest and control for extraneous variables, which may not be possible in field experiments. Hence, inferences drawn from laboratory experiments tend to be stronger in internal validity, but those from field experiments tend to be stronger in external validity. Experimental data is analyzed using quantitative statistical techniques. The primary strength of the experimental design is its strong internal validity due to its ability to isolate, control, and intensively examine a small number of variables, while its primary weakness is limited external generalizability since real life is often more complex (i.e., involve more extraneous variables) than contrived lab settings. Furthermore, if the research does not identify ex ante relevant extraneous variables and control for such variables, such lack of controls may hurt internal validity and may lead to spurious correlations.

Field surveys are non-experimental designs that do not control for or manipulate independent variables or treatments, but measure these variables and test their effects using statistical methods. Field surveys capture snapshots of practices, beliefs, or situations from a random sample of subjects in field settings through a survey questionnaire or less frequently, through a structured interview. In **cross-sectional field surveys**, independent and dependent variables are measured at the same point in time (e.g., using a single questionnaire), while in **longitudinal field surveys**, dependent variables are measured at a later point in time than the independent variables. The strengths of field surveys are their external validity (since data is collected in field settings), their ability to capture and control for a large number of variables, and their ability to study a problem from multiple perspectives or using multiple theories. However, because of their non-temporal nature, internal validity (cause-effect relationships) are difficult to infer, and surveys may be subject to respondent biases (e.g., subjects may provide a "socially desirable" response rather than their true response) which further hurts internal validity.

Secondary data analysis is an analysis of data that has previously been collected and tabulated by other sources. Such data may include data from government agencies such as employment statistics from the U.S. Bureau of Labor Services or development statistics by country from the United Nations Development Program, data collected by other researchers (often used in meta-analytic studies), or publicly available third-party data, such as financial data from stock markets or real-time auction data from eBay. This is in contrast to most other research designs where collecting primary data for research is part of the researcher's job. Secondary data analysis may be an effective means of research where primary data collection is too costly or infeasible, and secondary data is available at a level of analysis suitable for answering the researcher's questions. The limitations of this design are that the data might not have been collected in a systematic or scientific manner and hence unsuitable for scientific research, since the data was collected for a presumably different purpose, they may not adequately address the research questions of interest to the researcher, and interval validity is problematic if the temporal precedence between cause and effect is unclear.

Case research is an in-depth investigation of a problem in one or more real-life settings (case sites) over an extended period of time. Data may be collected using a combination of interviews, personal observations, and internal or external documents. Case studies can be positivist in nature (for hypotheses testing) or interpretive (for theory building). The strength of this research method is its ability to discover a wide variety of social, cultural, and political factors potentially related to the phenomenon of interest that may not be known in advance. Analysis tends to be qualitative in nature, but heavily contextualized and nuanced. However, interpretation of findings may depend on the observational and integrative ability of the researcher, lack of control may make it difficult to establish causality, and findings from a single case site may not be readily generalized to other case sites. Generalizability can be improved by replicating and comparing the analysis in other case sites in a **multiple case design**.

Focus group research is a type of research that involves bringing in a small group of subjects (typically 6 to 10 people) at one location, and having them discuss a phenomenon of interest for a period of 1.5 to 2 hours. The discussion is moderated and led by a trained facilitator, who sets the agenda and poses an initial set of questions for participants, makes sure that ideas and experiences of all participants are represented, and attempts to build a holistic understanding of the problem situation based on participants' comments and experiences. Internal validity cannot be established due to lack of controls and the findings may not be generalized to other settings because of small sample size. Hence, focus groups are not generally used for explanatory or descriptive research, but are more suited for exploratory research.

Action research assumes that complex social phenomena are best understood by introducing interventions or "actions" into those phenomena and observing the effects of those actions. In this method, the researcher is usually a consultant or an organizational member embedded within a social context such as an organization, who initiates an action such as new organizational procedures or new technologies, in response to a real problem such as declining profitability or operational bottlenecks. The researcher's choice of actions must be based on theory, which should explain why and how such actions may cause the desired change. The researcher then observes the results of that action, modifying it as necessary, while simultaneously learning from the action and generating theoretical insights about the target problem and interventions. The initial theory is validated by the extent to which the chosen action successfully solves the target problem. Simultaneous problem solving and insight generation is the central feature that distinguishes action research from all other research methods, and hence, action research is an excellent method for bridging research and practice. This method is also suited for studying unique social problems that cannot be replicated outside that context, but it is also subject to researcher bias and subjectivity, and the generalizability of findings is often restricted to the context where the study was conducted.

Ethnography is an interpretive research design inspired by anthropology that emphasizes that research phenomenon must be studied within the context of its culture. The researcher is deeply immersed in a certain culture over an extended period of time (8 months to 2 years), and during that period, engages, observes, and records the daily life of the studied culture, and theorizes about the evolution and behaviors in that culture. Data is collected primarily via observational techniques, formal and informal interaction with participants in that culture, and personal field notes, while data analysis involves "sense-making". The researcher must narrate her experience in great detail so that readers may experience that same culture without necessarily being there. The advantages of this approach are its sensitiveness to the context, the rich and nuanced understanding it generates, and minimal respondent bias. However, this is also an extremely time and resource-intensive approach, and findings are specific to a given culture and less generalizable to other cultures.

Selecting Research Designs

Given the above multitude of research designs, which design should researchers choose for their research? Generally speaking, researchers tend to select those research designs that they are most comfortable with and feel most competent to handle, but ideally, the choice should depend on the nature of the research phenomenon being studied. In the preliminary phases of research, when the research problem is unclear and the researcher wants to scope out the nature and extent of a certain research problem, a focus group (for individual unit of analysis) or a case study (for organizational unit of analysis) is an ideal strategy for exploratory research. As one delves further into the research domain, but finds that there are no good theories to explain the phenomenon of interest and wants to build a theory to fill in the unmet gap in that area, interpretive designs such as case research or ethnography may be useful designs. If competing theories exist and the researcher wishes to test these different theories or integrate them into a larger theory, positivist designs such as experimental design, survey research, or secondary data analysis are more appropriate.

Regardless of the specific research design chosen, the researcher should strive to collect quantitative and qualitative data using a combination of techniques such as questionnaires, interviews, observations, documents, or secondary data. For instance, even in a highly structured survey questionnaire, intended to collect quantitative data, the researcher may leave some room for a few open-ended questions to collect qualitative data that may generate unexpected insights not otherwise available from structured quantitative data alone. Likewise, while case research employ mostly face-to-face interviews to collect most qualitative data, the potential and value of collecting quantitative data should not be ignored. As an example, in a study of organizational decision making processes, the case interviewer can record numeric quantities such as how many months it took to make certain organizational decisions, how many people were involved in that decision process, and how many decision alternatives were considered, which can provide valuable insights not otherwise available from interviewees' narrative responses. Irrespective of the specific research design employed, the goal of the researcher should be to collect as much and as diverse data as possible that can help generate the best possible insights about the phenomenon of interest.

Chapter 6

Measurement of Constructs

Theoretical propositions consist of relationships between abstract constructs. Testing theories (i.e., theoretical propositions) require measuring these constructs accurately, correctly, and in a scientific manner, before the strength of their relationships can be tested. Measurement refers to careful, deliberate observations of the real world and is the essence of empirical research. While some constructs in social science research, such as a person's age, weight, or a firm's size, may be easy to measure, other constructs, such as creativity, prejudice, or alienation, may be considerably harder to measure. In this chapter, we will examine the related processes of conceptualization and operationalization for creating measures of such constructs.

Conceptualization

Conceptualization is the mental process by which fuzzy and imprecise constructs (concepts) and their constituent components are defined in concrete and precise terms. For instance, we often use the word "prejudice" and the word conjures a certain image in our mind; however, we may struggle if we were asked to define exactly what the term meant. If someone says bad things about other racial groups, is that racial prejudice? If women earn less than men for the same job, is that gender prejudice? If churchgoers believe that non-believers will burn in hell, is that religious prejudice? Are there different kinds of prejudice, and if so, what are they? Are there different levels of prejudice, such as high or low? Answering all of these questions is the key to measuring the prejudice construct correctly. The process of understanding what is included and what is excluded in the concept of prejudice is the conceptualization process.

The conceptualization process is all the more important because of the imprecision, vagueness, and ambiguity of many social science constructs. For instance, is "compassion" the same thing as "empathy" or "sentimentality"? If you have a proposition stating that "compassion is positively related to empathy", you cannot test that proposition unless you can conceptually separate empathy from compassion and then empirically measure these two very similar constructs correctly. If deeply religious people believe that some members of their society, such as nonbelievers, gays, and abortion doctors, will burn in hell for their sins, and forcefully try to change the "sinners" behaviors to prevent them from going to hell, are they acting in a prejudicial manner or a compassionate manner? Our definition of such constructs is not based on any objective criterion, but rather on a shared ("inter-subjective") agreement between our mental images (conceptions) of these constructs.

While defining constructs such as prejudice or compassion, we must understand that sometimes, these constructs are not real or can exist independently, but are simply imaginary creations in our mind. For instance, there may be certain tribes in the world who lack prejudice and who cannot even imagine what this concept entails. But in real life, we tend to treat this concept as real. The process of regarding mental constructs as real is called *reification*, which is central to defining constructs and identifying measurable variables for measuring them.

One important decision in conceptualizing constructs is specifying whether they are unidimensional and multidimensional. **Unidimensional** constructs are those that are expected to have a single underlying dimension. These constructs can be measured using a single measure or test. Examples include simple constructs such as a person's weight, wind speed, and probably even complex constructs like self-esteem (if we conceptualize self-esteem as consisting of a single dimension, which of course, may be a unrealistic assumption). **Multidimensional** constructs consist of two or more underlying dimensions. For instance, if we conceptualize a person's academic aptitude as consisting of two dimensions – mathematical and verbal ability – then academic aptitude is a multidimensional construct. Each of the underlying dimensions in this case must be measured separately, say, using different tests for mathematical and verbal ability, and the two scores can be combined, possibly in a weighted manner, to create an overall value for the academic aptitude construct.

Operationalization

Once a theoretical construct is defined, exactly how do we measure it? **Operationalization** refers to the process of developing **indicators** or items for measuring these constructs. For instance, if an unobservable theoretical construct such as socioeconomic status is defined as the level of family income, it can be operationalized using an indicator that asks respondents the question: what is your annual family income? Given the high level of subjectivity and imprecision inherent in social science constructs, we tend to measure most of those constructs (except a few demographic constructs such as age, gender, education, and income) using multiple indicators. This process allows us to examine the closeness amongst these indicators as an assessment of their accuracy (reliability).

Indicators operate at the empirical level, in contrast to constructs, which are conceptualized at the theoretical level. The combination of indicators at the empirical level representing a given construct is called a **variable**. As noted in a previous chapter, variables may be independent, dependent, mediating, or moderating, depending on how they are employed in a research study. Also each indicator may have several **attributes** (or levels) and each attribute represent a **value**. For instance, a "gender" variable may have two attributes: male or female. Likewise, a customer satisfaction scale may be constructed to represent five attributes: "strongly dissatisfied", "somewhat dissatisfied", "neutral", "somewhat satisfied" and "strongly satisfied". Values of attributes may be **quantitative** (numeric) or **qualitative** (non-numeric). Quantitative data can be analyzed using quantitative data analysis techniques, such as regression or structural equation modeling, while qualitative data require qualitative data analysis techniques, such as coding. Note that many variables in social science research are qualitative, even when represented in a quantitative manner. For instance, we can create a customer satisfaction indicator with five attributes: strongly dissatisfied, somewhat dissatisfied, neutral, somewhat satisfied, and strongly satisfied, and assign numbers 1 through 5 respectively for these five attributes, so that we can use sophisticated statistical tools for quantitative data analysis. However, note that the numbers are only labels associated with respondents' personal evaluation of their own satisfaction, and the underlying variable (satisfaction) is still qualitative even though we represented it in a quantitative manner.

Indicators may be reflective or formative. A **reflective indicator** is a measure that "reflects" an underlying construct. For example, if religiosity is defined as a construct that measures how religious a person is, then attending religious services may be a reflective indicator of religiosity. A **formative indicator** is a measure that "forms" or contributes to an underlying construct. Such indicators may represent different dimensions of the construct of interest. For instance, if religiosity is defined as composing of a belief dimension, a devotional dimension, and a ritual dimension, then indicators chosen to measure each of these different dimensions will be considered formative indicators. Unidimensional constructs are measured using reflective indicators (even though multiple reflective indicators may be used for measuring abstruse constructs such as self-esteem), while multidimensional constructs are measured as a formative combination of the multiple dimensions, even though each of the underlying dimensions may be measured using one or more reflective indicators.

Levels of Measurement

The first decision to be made in operationalizing a construct is to decide on what is the intended level of measurement. **Levels of measurement**, also called **rating scales**, refer to the values that an indicator can take (but says nothing about the indicator itself). For example, male and female (or M and F, or 1 and 2) are two levels of the indicator "gender." In his seminal article titled "On the theory of scales of measurement" published in *Science* in 1946, psychologist Stanley Smith Stevens (1946) defined four generic types of rating scales for scientific measurements: nominal, ordinal, interval, and ratio scales. The statistical properties of these scales are shown in Table 6.1.

Scale	Central Tendency	Statistics	Transformations
Nominal	Mode	Chi-square	One-to-one (equality)
Ordinal	Median	Percentile, non-parametric statistics	Monotonic increasing (order)
Interval	Arithmetic mean, range, standard deviation	Correlation, regression, analysis of variance	Positive linear (affine)
Ratio	Geometric mean, harmonic mean	Coefficient of variation	Positive similarities (multiplicative, logarithmic)
Note: All higher-order scales can use any of the statistics for lower order scales.			

Table 6.1. Statistical properties of rating scales

Nominal scales, also called categorical scales, measure categorical data. These scales are used for variables or indicators that have mutually exclusive attributes. Examples include gender (two values: male or female), industry type (manufacturing, financial, agriculture, etc.), and religious affiliation (Christian, Muslim, Jew, etc.). Even if we assign unique numbers to each value, for instance 1 for male and 2 for female, the numbers don't really mean anything (i.e., 1 is not less than or half of 2) and could have been easily been represented non-numerically, such as

M for male and F for female. Nominal scales merely offer *names* or *labels* for different attribute values. The appropriate measure of central tendency of a nominal scale is mode, and neither the mean nor the median can be defined. Permissible statistics are chi-square and frequency distribution, and only a one-to-one (equality) transformation is allowed (e.g., 1=Male, 2=Female).

Ordinal scales are those that measure *rank-ordered* data, such as the ranking of students in a class as first, second, third, and so forth, based on their grade point average or test scores. However, the actual or relative values of attributes or difference in attribute values cannot be assessed. For instance, ranking of students in class says nothing about the actual GPA or test scores of the students, or how they well performed relative to one another. A classic example in the natural sciences is Moh's scale of mineral hardness, which characterizes the hardness of various minerals by their ability to scratch other minerals. For instance, diamonds can scratch all other naturally occurring minerals on earth, and hence diamond is the "hardest" mineral. However, the scale does not indicate the actual hardness of these minerals or even provides a relative assessment of their hardness. Ordinal scales can also use attribute labels (anchors) such as "bad", "medium", and "good", or "strongly dissatisfied", "somewhat dissatisfied", "neutral", or "somewhat satisfied", and "strongly satisfied". In the latter case, we can say that respondents who are "somewhat satisfied" are less satisfied than those who are "strongly satisfied", but we cannot quantify their satisfaction levels. The central tendency measure of an ordinal scale can be its median or mode, and means are uninterpretable. Hence, statistical analyses may involve percentiles and non-parametric analysis, but more sophisticated techniques such as correlation, regression, and analysis of variance, are not appropriate. Monotonically increasing transformation (which retains the ranking) is allowed.

Interval scales are those where the values measured are not only rank-ordered, but are also equidistant from adjacent attributes. For example, the temperature scale (in Fahrenheit or Celsius), where the difference between 30 and 40 degree Fahrenheit is the same as that between 80 and 90 degree Fahrenheit. Likewise, if you have a scale that asks respondents' annual income using the following attributes (ranges): $0 to 10,000, $10,000 to 20,000, $20,000 to 30,000, and so forth, this is also an interval scale, because the mid-point of each range (i.e., $5,000, $15,000, $25,000, etc.) are equidistant from each other. The intelligence quotient (IQ) scale is also an interval scale, because the scale is designed such that the difference between IQ scores 100 and 110 is supposed to be the same as between 110 and 120 (although we do not really know whether that is truly the case). Interval scale allows us to examine "how much more" is one attribute when compared to another, which is not possible with nominal or ordinal scales. Allowed central tendency measures include mean, median, or mode, as are measures of dispersion, such as range and standard deviation. Permissible statistical analyses include all of those allowed for nominal and ordinal scales, plus correlation, regression, analysis of variance, and so on. Allowed scale transformation are positive linear. Note that the satisfaction scale discussed earlier is not strictly an interval scale, because we cannot say whether the difference between "strongly satisfied" and "somewhat satisfied" is the same as that between "neutral" and "somewhat satisfied" or between "somewhat dissatisfied" and "strongly dissatisfied". However, social science researchers often "pretend" (incorrectly) that these differences are equal so that we can use statistical techniques for analyzing ordinal scaled data.

Ratio scales are those that have all the qualities of nominal, ordinal, and interval scales, and in addition, also have a "true zero" point (where the value zero implies lack or non-availability of the underlying construct). Most measurement in the natural sciences and engineering, such as mass, incline of a plane, and electric charge, employ ratio scales, as are some social science variables such as age, tenure in an organization, and firm size (measured as employee count or gross revenues). For example, a firm of size zero means that it has no employees or revenues. The Kelvin temperature scale is also a ratio scale, in contrast to the Fahrenheit or Celsius scales, because the zero point on this scale (equaling -273.15 degree Celsius) is not an arbitrary value but represents a state where the particles of matter at this temperature have zero kinetic energy. These scales are called "ratio" scales because the ratios of two points on these measures are meaningful and interpretable. For example, a firm of size 10 employees is double that of a firm of size 5, and the same can be said for a firm of 10,000 employees relative to a different firm of 5,000 employees. All measures of central tendencies, including geometric and harmonic means, are allowed for ratio scales, as are ratio measures, such as studentized range or coefficient of variation. All statistical methods are allowed. Sophisticated transformation such as positive similar (e.g., multiplicative or logarithmic) are also allowed.

Based on the four generic types of scales discussed above, we can create specific rating scales for social science research. Common rating scales include binary, Likert, semantic differential, or Guttman scales. Other less common scales are not discussed here.

Binary scales. Binary scales are nominal scales consisting of binary items that assume one of two possible values, such as yes or no, true or false, and so on. For example, a typical binary scale for the "political activism" construct may consist of the six binary items shown in Table 6.2. Each item in this scale is a binary item, and the total number of "yes" indicated by a respondent (a value from 0 to 6) can be used as an overall measure of that person's political activism. To understand how these items were derived, refer to the "Scaling" section later on in this chapter. Binary scales can also employ other values, such as male or female for gender, full-time or part-time for employment status, and so forth. If an employment status item is modified to allow for more than two possible values (e.g., unemployed, full-time, part-time, and retired), it is no longer binary, but still remains a nominal scaled item.

Have you ever written a letter to a public official	Yes	No
Have you ever signed a political petition	Yes	No
Have you ever donated money to a political cause	Yes	No
Have you ever donated money to a candidate running for public office	Yes	No
Have you ever written a political letter to the editor of a newspaper or magazine	Yes	No
Have you ever persuaded someone to change his/her voting plans	Yes	No

Table 6.2. A six-item binary scale for measuring political activism

Likert scale. Designed by Rensis Likert, this is a very popular rating scale for measuring ordinal data in social science research. This scale includes Likert items that are simply-worded statements to which respondents can indicate their extent of agreement or disagreement on a five or seven-point scale ranging from "strongly disagree" to "strongly agree". A typical example of a six-item Likert scale for the "employment self-esteem" construct is shown in Table 6.3. Likert scales are summated scales, that is, the overall scale score may be a summation of the attribute values of each item as selected by a respondent.

	Strongly Disagree	Somewhat Disagree	Neutral	Somewhat Agree	Strongly Agree
I feel good about my job	1	2	3	4	5
I get along well with others at work	1	2	3	4	5
I'm proud of my relationship with my supervisor at work	1	2	3	4	5
I can tell that other people at work are glad to have me there	1	2	3	4	5
I can tell that my coworkers respect me	1	2	3	4	5
I feel like I make a useful contribution at work	1	2	3	4	5

Table 6.3. A six-item Likert scale for measuring employment self-esteem

Likert items allow for more granularity (more finely tuned response) than binary items, including whether respondents are neutral to the statement. Three or nine values (often called "anchors") may also be used, but it is important to use an odd number of values to allow for a "neutral" (or "neither agree nor disagree") anchor. Some studies have used a "forced choice approach" to force respondents to agree or disagree with the LIkert statement by dropping the neutral mid-point and using even number of values and, but this is not a good strategy because some people may indeed be neutral to a given statement and the forced choice approach does not provide them the opportunity to record their neutral stance. A key characteristic of a Likert scale is that even though the statements vary in different items or indicators, the anchors ("strongly disagree" to "strongly agree") remain the same. Likert scales are ordinal scales because the anchors are not necessarily equidistant, even though sometimes we treat them like interval scales.

How would you rate your opinions on national health insurance?						
	Very much	Somewhat	Neither	Somewhat	Very much	
Good	□	□	□	□	□	Bad
Useful	□	□	□	□	□	Useless
Caring	□	□	□	□	□	Uncaring
Interesting	□	□	□	□	□	Boring

Table 6.4. A semantic differential scale for measuring attitude toward national health insurance

Semantic differential scale. This is a composite (multi-item) scale where respondents are asked to indicate their opinions or feelings toward a single statement using different pairs of adjectives framed as polar opposites. For instance, the construct "attitude toward national health insurance" can be measured using four items shown in Table 6.4. As in the Likert scale, the overall scale score may be a summation of individual item scores. Notice that in Likert scales, the statement changes but the anchors remain the same across items. However, in semantic differential scales, the statement remains constant, while the anchors (adjective pairs) change across items. Semantic differential is believed to be an excellent technique for measuring people's attitude or feelings toward objects, events, or behaviors.

Guttman scale. Designed by Louis Guttman, this composite scale uses a series of items arranged in increasing order of intensity of the construct of interest, from least intense to most intense. As an example, the construct "attitude toward immigrants" can be measured using five items shown in Table 6.5. Each item in the above Guttman scale has a weight (not indicated above) which varies with the intensity of that item, and the weighted combination of each response is used as aggregate measure of an observation.

How will you rate your opinions on the following statements about immigrants?		
Do you mind immigrants being citizens of your country	Yes	No
Do you mind immigrants living in your own neighborhood	Yes	No
Would you mind living next door to an immigrant	Yes	No
Would you mind having an immigrant as your close friend	Yes	No
Would you mind if someone in your family married an immigrant	Yes	No

Table 6.5. A five-item Guttman scale for measuring attitude toward immigrants

Scaling

The previous section discussed how to measure respondents' responses to predesigned items or indicators belonging to an underlying construct. But how do we create the indicators themselves? The process of creating the indicators is called scaling. More formally, **scaling** is a branch of measurement that involves the construction of measures by associating qualitative judgments about unobservable constructs with quantitative, measurable metric units. Stevens (1946) said, "Scaling is the assignment of objects to numbers according to a rule." This process of measuring abstract concepts in concrete terms remains one of the most difficult tasks in empirical social science research.

The outcome of a scaling process is a **scale**, which is an empirical structure for measuring items or indicators of a given construct. Understand that "scales", as discussed in this section, are a little different from "rating scales" discussed in the previous section. A rating scale is used to capture the respondents' reactions to a given item, for instance, such as a nominal scaled item captures a yes/no reaction and an interval scaled item captures a value between "strongly disagree" to "strongly agree." Attaching a rating scale to a statement or instrument is not scaling. Rather, scaling is the formal process of developing scale items, before rating scales can be attached to those items.

Scales can be unidimensional or multidimensional, based on whether the underlying construct is unidimensional (e.g., weight, wind speed, firm size) or multidimensional (e.g., academic aptitude, intelligence). Unidimensional scale measures constructs along a single scale, ranging from high to low. Note that some of these scales may include multiple items, but all of these items attempt to measure the same underlying dimension. This is particularly the case with many social science constructs such as self-esteem, which are assumed to have a single dimension going from low to high. Multi-dimensional scales, on the other hand, employ different items or tests to measure each dimension of the construct separately, and then combine the scores on each dimension to create an overall measure of the multidimensional construct. For instance, academic aptitude can be measured using two separate tests of students' mathematical and verbal ability, and then combining these scores to create an overall measure for academic aptitude. Since most scales employed in social science research are unidimensional, we will next three examine approaches for creating unidimensional scales.

Unidimensional scaling methods were developed during the first half of the twentieth century and were named after their creators. The three most popular unidimensional scaling methods are: (1) Thurstone's equal-appearing scaling, (2) Likert's summative scaling, and (3) Guttman's cumulative scaling. The three approaches are similar in many respects, with the key differences being the rating of the scale items by judges and the statistical methods used to select the final items. Each of these methods are discussed next.

Thurstone's equal-appearing scaling method. Louis Thurstone. one of the earliest and most famous scaling theorists, published a method of equal-appearing intervals in 1925. This method starts with a clear conceptual definition of the construct of interest. Based on this definition, potential scale items are generated to measure this construct. These items are generated by experts who know something about the construct being measured. The initial pool of candidate items (ideally 80 to 100 items) should be worded in a similar manner, for instance, by framing them as statements to which respondents may agree or disagree (and not as questions or other things). Next, a panel of judges is recruited to select specific items from this candidate pool to represent the construct of interest. Judges may include academics trained in the process of instrument construction or a random sample of respondents of interest (i.e., people who are familiar with the phenomenon). The selection process is done by having each judge independently rate each item on a scale from 1 to 11 based on how closely, in their opinion, that item reflects the intended construct (1 represents extremely unfavorable and 11 represents extremely favorable). For each item, compute the median and inter-quartile range (the difference between the 75th and the 25th percentile – a measure of dispersion), which are plotted on a histogram, as shown in Figure 6.1. The final scale items are selected as statements that are at equal intervals across a range of medians. This can be done by grouping items with a common median, and then selecting the item with the smallest inter-quartile range within each median group. However, instead of relying entirely on statistical analysis for item selection, a better strategy may be to examine the candidate items at each level and selecting the statement that is the most clear and makes the most sense. The median value of each scale item represents the weight to be used for aggregating the items into a composite scale score representing the construct of interest. We now have a scale which looks like a ruler, with one item or statement at each of the 11 points on the ruler (and weighted as such). Because items appear equally throughout the entire 11-pointrange of the scale, this technique is called an equal-appearing scale.

Thurstone also created two additional methods of building unidimensional scales – the *method of successive intervals* and the *method of paired comparisons* – which are both very similar to the method of equal-appearing intervals, except for how judges are asked to rate the data. For instance, the method of paired comparison requires each judge to make a judgment between each pair of statements (rather than rate each statement independently on a 1 to 11 scale). Hence, the name paired comparison method. With a lot of statements, this approach can be enormously time consuming and unwieldy compared to the method of equal-appearing intervals.

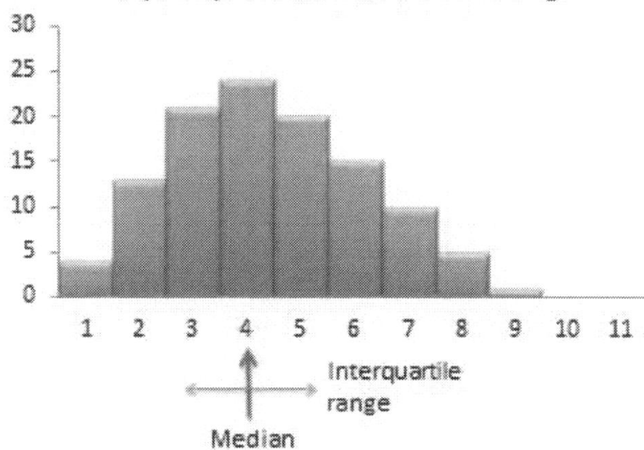

Figure 6.1. Histogram for Thurstone scale items

Likert's summative scaling method. The Likert method, a unidimensional scaling method developed by Murphy and Likert (1938), is quite possibly the most popular of the three scaling approaches described in this chapter. As with Thurstone's method, the Likert method also starts with a clear definition of the construct of interest, and using a set of experts to generate about 80 to 100 potential scale items. These items are then rated by judges on a 1 to 5 (or 1 to 7) rating scale as follows: 1 for strongly disagree with the concept, 2 for somewhat disagree with the concept, 3 for undecided, 4 for somewhat agree with the concept, and 5 for strongly agree with the concept. Following this rating, specific items can be selected for the final scale can be selected in one of several ways: (1) by computing bivariate correlations between judges rating of each item and the total item (created by summing all individual items for each respondent), and throwing out items with low (e.g., less than 0.60) item-to-total correlations, or (2) by averaging the rating for each item for the top quartile and the bottom quartile of judges, doing a t-test for the difference in means, and selecting items that have high t-values (i.e., those that discriminates best between the top and bottom quartile responses). In the end, researcher's judgment may be used to obtain a relatively small (say 10 to 15) set of items that have high item-to-total correlations and high discrimination (i.e., high t-values). The Likert method assumes equal weights for all items, and hence, respondent's responses to each item can be summed to create a composite score for that respondent. Hence, this method is called a summated scale. Note that any item with reversed meaning from the original direction of the construct must be reverse coded (i.e., 1 becomes a 5, 2 becomes a 4, and so forth) before summating.

Guttman's cumulative scaling method. Designed by Guttman (1950), the cumulative scaling method is based on Emory Bogardus' social distance technique, which assumes that people's willingness to participate in social relations with other people vary in degrees of intensity, and measures that intensity using a list of items arranged from "least intense" to "most intense". The idea is that people who agree with one item on this list also agree with all previous items. In practice, we seldom find a set of items that matches this cumulative pattern perfectly. A scalogram analysis is used to examine how closely a set of items corresponds to the idea of cumulativeness.

Like previous scaling methods, the Guttman method also starts with a clear definition of the construct of interest, and then using experts to develop a large set of candidate items. A group of judges then rate each candidate item as "yes" if they view the item as being favorable to the construct and "no" if they see the item as unfavorable. Next, a matrix or table is created showing the judges' responses to all candidate items. This matrix is sorted in decreasing order from judges with more "yes" at the top to those with fewer "yes" at the bottom. Judges with the same number of "yes", the statements can be sorted from left to right based on most number of agreements to least. The resulting matrix will resemble Table 6.6. Notice that the scale is now almost cumulative when read from left to right (across the items). However, there may be a few exceptions, as shown in Table 6.6, and hence the scale is not entirely cumulative. To determine a set of items that best approximates the cumulativeness property, a data analysis technique called scalogram analysis can be used (or this can be done visually if the number of items is small). The statistical technique also estimates a score for each item that can be used to compute a respondent's overall score on the entire set of items.

Respondent	Item 12	Item 5	Item 3	Item 22	Item 8	Item 7	...
29	Y	Y	Y	Y	Y	Y	
7	Y	Y	Y	-	**Y**	-	
15	Y	Y	Y	Y	-	-	
3	Y	Y	Y	Y	-	-	
32	Y	Y	Y	-	-	-	
4	Y	Y	-	**Y**	-	-	
5	Y	Y	-	-	-	-	
23	Y	Y	-	-	-	-	
11	Y	-	-	**Y**	-	-	
Y indicates exceptions that prevents this matrix from being perfectly cumulative							

Table 6.6. Sorted rating matrix for a Guttman scale

Indexes

An **index** is a composite score derived from aggregating measures of *multiple constructs* (called components) using a set of rules and formulas. It is different from scales in that scales also aggregate measures, but these measures measure different dimensions or the same dimension of a *single construct*. A well-known example of an index is the consumer price index (CPI), which is computed every month by the Bureau of Labor Statistics of the U.S. Department of Labor. The CPI is a measure of how much consumers have to pay for goods and services in general, and is divided into eight major categories (food and beverages, housing, apparel, transportation, healthcare, recreation, education and communication, and "other goods and services"), which are further subdivided into more than 200 smaller items. Each month, government employees call all over the country to get the current prices of more than 80,000 items. Using a complicated weighting scheme that takes into account the location and probability of purchase of each item, these prices are combined by analysts, which are then combined into an overall index score using a series of formulas and rules.

Another example of index is socio-economic status (SES), also called the Duncan socioeconomic index (SEI). This index is a combination of three constructs: income, education, and occupation. Income is measured in dollars, education in years or degrees achieved, and occupation is classified into categories or levels by status. These very different measures are combined to create an overall SES index score, using a weighted combination of "occupational education" (percentage of people in that occupation who had one or more year of college education) and "occupational income" (percentage of people in that occupation who earned more than a specific annual income). However, SES index measurement has generated a lot of controversy and disagreement among researchers.

The process of creating an index is similar to that of a scale. First, conceptualize (define) the index and its constituent components. Though this appears simple, there may be a lot of disagreement among judges on what components (constructs) should be included or excluded from an index. For instance, in the SES index, isn't income correlated with education and occupation, and if so, should we include one component only or all three components? Reviewing the literature, using theories, and/or interviewing experts or key stakeholders may help resolve this issue. Second, operationalize and measure each component. For instance, how will you categorize occupations, particularly since some occupations may have changed with time (e.g., there were no Web developers before the Internet). Third, create a rule or formula for calculating the index score. Again, this process may involve a lot of subjectivity. Lastly, validate the index score using existing or new data.

Though indexes and scales yield a single numerical score or value representing a construct of interest, they are different in many ways. First, indexes often comprise of components that are very different from each other (e.g., income, education, and occupation in the SES index) and are measured in different ways. However, scales typically involve a set of similar items that use the same rating scale (such as a five-point Likert scale). Second, indexes often combine objectively measurable values such as prices or income, while scales are designed to assess subjective or judgmental constructs such as attitude, prejudice, or self-esteem. Some argue that the sophistication of the scaling methodology makes scales different from indexes, while others suggest that indexing methodology can be equally sophisticated. Nevertheless, indexes and scales are both essential tools in social science research.

Typologies

Scales and indexes generate ordinal measures of unidimensional constructs. However, researchers sometimes wish to summarize measures of two or more constructs to create a set of categories or types called a **typology**. Unlike scales or indexes, typologies are multi-dimensional but include only nominal variables. For instance, one can create a political typology of newspapers based on their orientation toward domestic and foreign policy, as expressed in their editorial columns, as shown in Figure 6.2. This typology can be used to categorize newspapers into one of four "ideal types" (A through D), identify the distribution of newspapers across these ideal types, and perhaps even create a classificatory model to classifying newspapers into one of these four ideal types depending on other attributes.

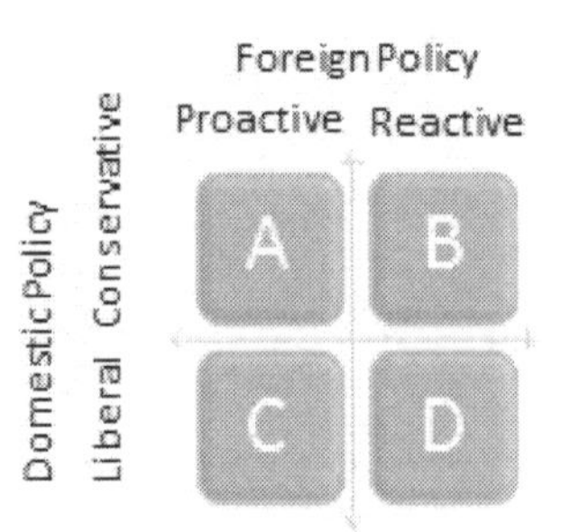

Figure 6.2. A multi-dimensional typology of newspapers

Summary

In closing, scale (or index) construction in social science research is a complex process involving several key decisions. Some of these decisions are:

- Should you use a scale, index, or typology?
- How do you plan to analyze the data?
- What is your desired level of measurement (nominal, ordinal, interval, or ratio) or rating scale?
- How many scale attributes should you use (e.g., 1 to 10; 1 to 7; −3 to +3)?
- Should you use an odd or even number of attributes (i.e., do you wish to have neutral or mid-point value)?
- How do you wish to label the scale attributes (especially for semantic differential scales)?
- Finally, what procedure would you use to generate the scale items (e.g., Thurstone, Likert, or Guttman method) or index components?

This chapter examined the process and outcomes of scale development. The next chapter will examine how to evaluate the reliability and validity of the scales developed using the above approaches.

Chapter 7

Scale Reliability and Validity

The previous chapter examined some of the difficulties with measuring constructs in social science research. For instance, how do we know whether we are measuring "compassion" and not the "empathy", since both constructs are somewhat similar in meaning? Or is compassion the same thing as empathy? What makes it more complex is that sometimes these constructs are imaginary concepts (i.e., they don't exist in reality), and multi-dimensional (in which case, we have the added problem of identifying their constituent dimensions). Hence, it is not adequate just to measure social science constructs using any scale that we prefer. We also must test these scales to ensure that: (1) these scales indeed measure the unobservable construct that we wanted to measure (i.e., the scales are "valid"), and (2) they measure the intended construct consistently and precisely (i.e., the scales are "reliable"). Reliability and validity, jointly called the "psychometric properties" of measurement scales, are the yardsticks against which the adequacy and accuracy of our measurement procedures are evaluated in scientific research.

A measure can be reliable but not valid, if it is measuring something very consistently but is consistently measuring the wrong construct. Likewise, a measure can be valid but not reliable if it is measuring the right construct, but not doing so in a consistent manner. Using the analogy of a shooting target, as shown in Figure 7.1, a multiple-item measure of a construct that is both reliable and valid consists of shots that clustered within a narrow range near the center of the target. A measure that is valid but not reliable will consist of shots centered on the target but not clustered within a narrow range, but rather scattered around the target. Finally, a measure that is reliable but not valid will consist of shots clustered within a narrow range but off from the target. Hence, reliability and validity are both needed to assure adequate measurement of the constructs of interest.

Figure 7.1. Comparison of reliability and validity

Reliability

Reliability is the degree to which the measure of a construct is consistent or dependable. In other words, if we use this scale to measure the same construct multiple times, do we get pretty much the same result every time, assuming the underlying phenomenon is not changing? An example of an unreliable measurement is people guessing your weight. Quite likely, people will guess differently, the different measures will be inconsistent, and therefore, the "guessing" technique of measurement is unreliable. A more reliable measurement may be to use a weight scale, where you are likely to get the same value every time you step on the scale, unless your weight has actually changed between measurements.

Note that reliability implies consistency but not accuracy. In the previous example of the weight scale, if the weight scale is calibrated incorrectly (say, to shave off ten pounds from your true weight, just to make you feel better!), it will not measure your true weight and is therefore not a valid measure. Nevertheless, the miscalibrated weight scale will still give you the same weight every time (which is ten pounds less than your true weight), and hence the scale is reliable.

What are the sources of unreliable observations in social science measurements? One of the primary sources is the observer's (or researcher's) subjectivity. If employee morale in a firm is measured by watching whether the employees smile at each other, whether they make jokes, and so forth, then different observers may infer different measures of morale if they are watching the employees on a very busy day (when they have no time to joke or chat) or a light day (when they are more jovial or chatty). Two observers may also infer different levels of morale on the same day, depending on what they view as a joke and what is not. "Observation" is a qualitative measurement technique. Sometimes, reliability may be improved by using quantitative measures, for instance, by counting the number of grievances filed over one month as a measure of (the inverse of) morale. Of course, grievances may or may not be a valid measure of morale, but it is less subject to human subjectivity, and therefore more reliable. A second source of unreliable observation is asking imprecise or ambiguous questions. For instance, if you ask people what their salary is, different respondents may interpret this question differently as monthly salary, annual salary, or per hour wage, and hence, the resulting observations will likely be highly divergent and unreliable. A third source of unreliability is asking questions about issues that respondents are not very familiar about or care about, such as asking an American college graduate whether he/she is satisfied with Canada's relationship with Slovenia, or asking a Chief Executive Officer to rate the effectiveness of his company's technology strategy – something that he has likely delegated to a technology executive.

So how can you create reliable measures? If your measurement involves soliciting information from others, as is the case with much of social science research, then you can start by replacing data collection techniques that depends more on researcher subjectivity (such as observations) with those that are less dependent on subjectivity (such as questionnaire), by asking only those questions that respondents may know the answer to or issues that they care about, by avoiding ambiguous items in your measures (e.g., by clearly stating whether you are looking for annual salary), and by simplifying the wording in your indicators so that they not misinterpreted by some respondents (e.g., by avoiding difficult words whose meanings they may not know). These strategies can improve the reliability of our measures, even though they will not necessarily make the measurements completely reliable. Measurement instruments must still be tested for reliability. There are many ways of estimating reliability, which are discussed next.

Inter-rater reliability. Inter-rater reliability, also called inter-observer reliability, is a measure of consistency between two or more independent raters (observers) of the same construct. Usually, this is assessed in a pilot study, and can be done in two ways, depending on the level of measurement of the construct. If the measure is categorical, a set of all categories is defined, raters check off which category each observation falls in, and the percentage of agreement between the raters is an estimate of inter-rater reliability. For instance, if there are two raters rating 100 observations into one of three possible categories, and their ratings match for 75% of the observations, then inter-rater reliability is 0.75. If the measure is interval or ratio scaled (e.g., classroom activity is being measured once every 5 minutes by two raters on 1 to 7 response scale), then a simple correlation between measures from the two raters can also serve as an estimate of inter-rater reliability.

Test-retest reliability. Test-retest reliability is a measure of consistency between two measurements (tests) of the same construct administered to the same sample at two different points in time. If the observations have not changed substantially between the two tests, then the measure is reliable. The correlation in observations between the two tests is an estimate of test-retest reliability. Note here that the time interval between the two tests is critical. Generally, the longer is the time gap, the greater is the chance that the two observations may change during this time (due to random error), and the lower will be the test-retest reliability.

Split-half reliability. Split-half reliability is a measure of consistency between two halves of a construct measure. For instance, if you have a ten-item measure of a given construct, randomly split those ten items into two sets of five (unequal halves are allowed if the total number of items is odd), and administer the entire instrument to a sample of respondents. Then, calculate the total score for each half for each respondent, and the correlation between the total scores in each half is a measure of split-half reliability. The longer is the instrument, the more likely it is that the two halves of the measure will be similar (since random errors are minimized as more items are added), and hence, this technique tends to systematically overestimate the reliability of longer instruments.

Internal consistency reliability. Internal consistency reliability is a measure of consistency between different items of the same construct. If a multiple-item construct measure is administered to respondents, the extent to which respondents rate those items in a similar manner is a reflection of internal consistency. This reliability can be estimated in terms of average inter-item correlation, average item-to-total correlation, or more commonly, Cronbach's alpha. As an example, if you have a scale with six items, you will have fifteen different item pairings, and fifteen correlations between these six items. Average inter-item correlation is the average of these fifteen correlations. To calculate average item-to-total correlation, you have to first create a "total" item by adding the values of all six items, compute the correlations between this total item and each of the six individual items, and finally, average the six correlations. Neither of the two above measures takes into account the number of items in the measure (six items in this example). Cronbach's alpha, a reliability measure designed by Lee Cronbach in 1951, factors in scale size in reliability estimation, calculated using the following formula:

$$\alpha = \frac{K}{K-1}\left(1 - \frac{\sum_{i=1}^{K}\sigma_{Y_i}^2}{\sigma_X^2}\right)$$

where K is the number of items in the measure, σ_X^2 is the variance (square of standard deviation) of the observed total scores, and $\sigma_{Y_i}^2$ is the observed variance for item i. The standardized Cronbach's alpha can be computed using a simpler formula:

$$\alpha_{\text{standardized}} = \frac{K\bar{r}}{(1 + (K-1)\bar{r})}$$

where K is the number of items, $\bar{r}$ is the average inter-item correlation, i.e., the mean of $K(K-1)/2$ coefficients in the upper triangular (or lower triangular) correlation matrix.

Validity

Validity, often called construct validity, refers to the extent to which a measure adequately represents the underlying construct that it is supposed to measure. For instance, is a measure of compassion really measuring compassion, and not measuring a different construct such as empathy? Validity can be assessed using theoretical or empirical approaches, and should ideally be measured using both approaches. Theoretical assessment of validity focuses on how well the idea of a theoretical construct is translated into or represented in an operational measure. This type of validity is called **translational validity** (or representational validity), and consists of two subtypes: face and content validity. Translational validity is typically assessed using a panel of expert judges, who rate each item (indicator) on how well they fit the conceptual definition of that construct, and a qualitative technique called Q-sort.

Empirical assessment of validity examines how well a given measure relates to one or more external criterion, based on empirical observations. This type of validity is called **criterion-related validity**, which includes four sub-types: convergent, discriminant, concurrent, and predictive validity. While translation validity examines whether a measure is a good reflection of its underlying construct, criterion-related validity examines whether a given measure behaves the way it should, given the theory of that construct. This assessment is based on quantitative analysis of observed data using statistical techniques such as correlational analysis, factor analysis, and so forth. The distinction between theoretical and empirical assessment of validity is illustrated in Figure 7.2. However, both approaches are needed to adequately ensure the validity of measures in social science research.

Note that the different types of validity discussed here refer to the validity of the *measurement procedures*, which is distinct from the validity of *hypotheses testing procedures*, such as internal validity (causality), external validity (generalizability), or statistical conclusion validity. The latter types of validity are discussed in a later chapter.

Face validity. Face validity refers to whether an indicator seems to be a reasonable measure of its underlying construct "on its face". For instance, the frequency of one's attendance at religious services seems to make sense as an indication of a person's religiosity without a lot of explanation. Hence this indicator has face validity. However, if we were to suggest how many books were checked out of an office library as a measure of employee morale, then such a measure would probably lack face validity because it does not seem to make much sense. Interestingly, some of the popular measures used in organizational research appears to lack face validity. For instance, absorptive capacity of an organization (how much new knowledge can it assimilate for improving organizational processes) has often been measured as research and development intensity (i.e., R&D expenses divided by gross revenues)! If your research includes constructs that are highly abstract or constructs that are hard to conceptually separate from each other (e.g., compassion and empathy), it may be worthwhile to consider using a panel of experts to evaluate the face validity of your construct measures.

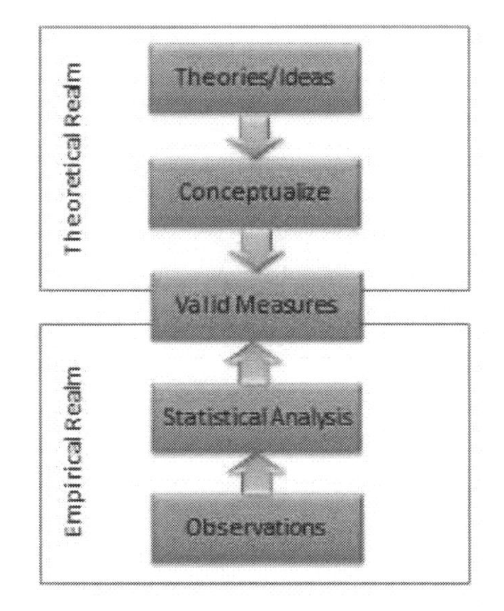

Figure 7.2. Two approaches of validity assessment

Content validity. Content validity is an assessment of how well a set of scale items matches with the relevant content domain of the construct that it is trying to measure. For instance, if you want to measure the construct "satisfaction with restaurant service," and you define the content domain of restaurant service as including the quality of food, courtesy of wait staff, duration of wait, and the overall ambience of the restaurant (i.e., whether it is noisy, smoky, etc.), then for adequate content validity, this construct should be measured using indicators that examine the extent to which a restaurant patron is satisfied with the quality of food, courtesy of wait staff, the length of wait, and the restaurant's ambience. Of course, this approach requires a detailed description of the entire content domain of a construct, which may be difficult for complex constructs such as self-esteem or intelligence. Hence, it may not be always possible to adequately assess content validity. As with face validity, an expert panel of judges may be employed to examine content validity of constructs.

Convergent validity refers to the closeness with which a measure relates to (or converges on) the construct that it is purported to measure, and **discriminant validity** refers to the degree to which a measure does not measure (or discriminates from) other constructs that it is not supposed to measure. Usually, convergent validity and discriminant validity are assessed jointly for a set of related constructs. For instance, if you expect that an organization's knowledge is related to its performance, how can you assure that your measure of organizational knowledge is indeed measuring organizational knowledge (for convergent validity) and not organizational performance (for discriminant validity)? Convergent validity can be established by comparing the observed values of one indicator of one construct with that of other indicators of the same construct and demonstrating similarity (or high correlation) between values of these indicators. Discriminant validity is established by demonstrating that indicators of one construct are dissimilar from (i.e., have low correlation with) other constructs. In the above example, if we have a three-item measure of organizational knowledge and three more items for organizational performance, based on observed sample data, we can compute bivariate correlations between each pair of knowledge and performance items. If this correlation matrix shows high correlations within items of the organizational knowledge and organizational performance constructs, but low correlations between items of these constructs, then we have simultaneously demonstrated convergent and discriminant validity (see Table 7.1).

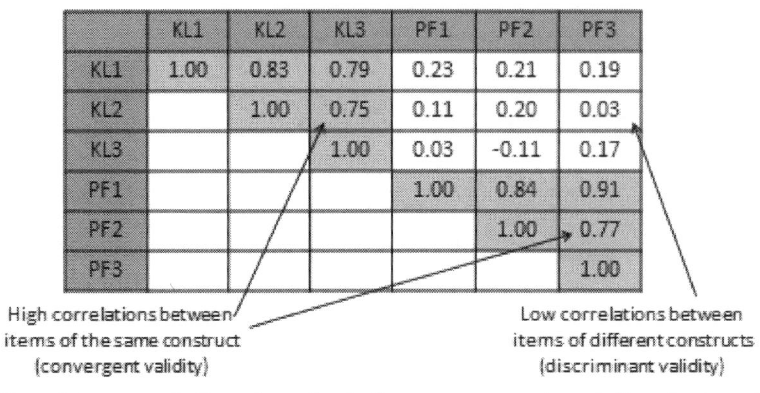

	KL1	KL2	KL3	PF1	PF2	PF3
KL1	1.00	0.83	0.79	0.23	0.21	0.19
KL2		1.00	0.75	0.11	0.20	0.03
KL3			1.00	0.03	-0.11	0.17
PF1				1.00	0.84	0.91
PF2					1.00	0.77
PF3						1.00

Table 7.1. Bivariate correlational analysis for convergent and discriminant validity

An alternative and more common statistical method used to demonstrate convergent and discriminant validity is *exploratory factor analysis*. This is a data reduction technique which aggregates a given set of items to a smaller set of factors based on the bivariate correlation structure discussed above using a statistical technique called principal components analysis. These factors should ideally correspond to the underling theoretical constructs that we are trying to measure. The general norm for factor extraction is that each extracted factor should have an eigenvalue greater than 1.0. The extracted factors can then be rotated using orthogonal or oblique rotation techniques, depending on whether the underlying constructs are expected to be relatively uncorrelated or correlated, to generate factor weights that can be used to aggregate the individual items of each construct into a composite measure. For adequate convergent validity, it is expected that items belonging to a common construct should exhibit factor loadings of 0.60 or higher on a single factor (called same-factor loadings), while for discriminant validity, these items should have factor loadings of 0.30 or less on all other factors (cross-factor loadings), as shown in rotated factor matrix example in Table 7.2. A more sophisticated technique for evaluating convergent and discriminant validity is the multi-trait multi-method (MTMM) approach. This technique requires measuring each construct (trait) using two or more different methods (e.g., survey and personal observation, or perhaps survey of two different respondent groups such as teachers and parents for evaluating academic quality). This is an onerous and relatively less popular approach, and is therefore not discussed here.

Criterion-related validity can also be assessed based on whether a given measure relate well with a current or future criterion, which are respectively called concurrent and predictive validity. **Predictive validity** is the degree to which a measure successfully predicts a future outcome that it is theoretically expected to predict. For instance, can standardized test scores (e.g., Scholastic Aptitude Test scores) correctly predict the academic success in college (e.g., as measured by college grade point average)? Assessing such validity requires creation of a "nomological network" showing how constructs are theoretically related to each other. **Concurrent validity** examines how well one measure relates to other concrete criterion that is presumed to occur simultaneously. For instance, do students' scores in a calculus class correlate well with their scores in a linear algebra class? These scores should be related concurrently because they are both tests of mathematics. Unlike convergent and discriminant validity, concurrent and predictive validity is frequently ignored in empirical social science research.

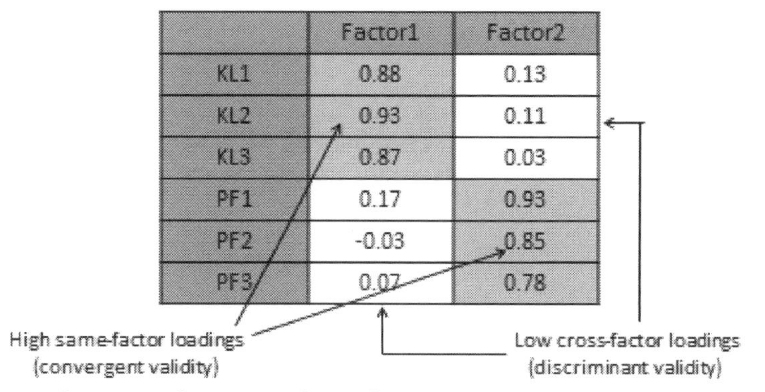

	Factor1	Factor2
KL1	0.88	0.13
KL2	0.93	0.11
KL3	0.87	0.03
PF1	0.17	0.93
PF2	-0.03	0.85
PF3	0.07	0.78

Table 7.2. Exploratory factor analysis for convergent and discriminant validity

Theory of Measurement

Now that we know the different kinds of reliability and validity, let us try to synthesize our understanding of reliability and validity in a mathematical manner using *classical test theory*, also called *true score theory*. This is a psychometric theory that examines how measurement works, what it measures, and what it does not measure. This theory postulates that every observation has a *true score* T that can be observed accurately if there were no errors in measurement. However, the presence of *measurement errors* E results in a deviation of the *observed score* X from the true score as follows:

$$\underset{\text{Observed score}}{X} = \underset{\text{True score}}{T} + \underset{\text{Error}}{E}$$

Across a set of observed scores, the variance of observed and true scores can be related using a similar equation:

$$\text{var}(X) = \text{var}(T) + \text{var}(E)$$

The goal of psychometric analysis is to estimate and minimize if possible the error variance var(E), so that the observed score X is a good measure of the true score T.

Measurement errors can be of two types: random error and systematic error. **Random error** is the error that can be attributed to a set of unknown and uncontrollable external factors that randomly influence some observations but not others. As an example, during the time of measurement, some respondents may be in a nicer mood than others, which may influence how they respond to the measurement items. For instance, respondents in a nicer mood may respond more positively to constructs like self-esteem, satisfaction, and happiness than those who are in a poor mood. However, it is not possible to anticipate which subject is in what type of mood or control for the effect of mood in research studies. Likewise, at an organizational level, if we are measuring firm performance, regulatory or environmental changes may affect the performance of some firms in an observed sample but not others. Hence, random error is considered to be "noise" in measurement and generally ignored.

Systematic error is an error that is introduced by factors that systematically affect all observations of a construct across an entire sample in a systematic manner. In our previous example of firm performance, since the recent financial crisis impacted the performance of financial firms disproportionately more than any other type of firms such as manufacturing or service firms, if our sample consisted only of financial firms, we may expect a systematic reduction in performance of all firms in our sample due to the financial crisis. Unlike random error, which may be positive negative, or zero, across observation in a sample, systematic errors tends to be consistently positive or negative across the entire sample. Hence, systematic error is sometimes considered to be "bias" in measurement and should be corrected.

Since an observed score may include both random and systematic errors, our true score equation can be modified as:

$$X = T + E_r + E_s$$

where E_r and E_s represent random and systematic errors respectively. The statistical impact of these errors is that random error adds variability (e.g., standard deviation) to the distribution of an observed measure, but does not affect its central tendency (e.g., mean), while systematic error affects the central tendency but not the variability, as shown in Figure 7.3.

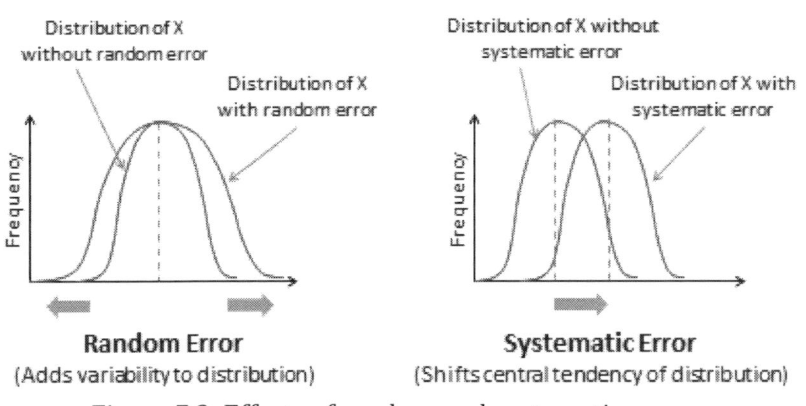

Figure 7.3. Effects of random and systematic errors

What does random and systematic error imply for measurement procedures? By increasing variability in observations, random error reduces the reliability of measurement. In contrast, by shifting the central tendency measure, systematic error reduces the validity of measurement. Validity concerns are far more serious problems in measurement than reliability concerns, because an invalid measure is probably measuring a different construct than what we intended, and hence validity problems cast serious doubts on findings derived from statistical analysis.

Note that reliability is a ratio or a fraction that captures how close the true score is relative to the observed score. Hence, reliability can be expressed as:

$$var(T) / var(X) = var(T) / [var(T) + var(E)]$$

If var(T) = var(X), then the true score has the same variability as the observed score, and the reliability is 1.0.

An Integrated Approach to Measurement Validation

A complete and adequate assessment of validity must include both theoretical and empirical approaches. As shown in Figure 7.4, this is an elaborate multi-step process that must take into account the different types of scale reliability and validity.

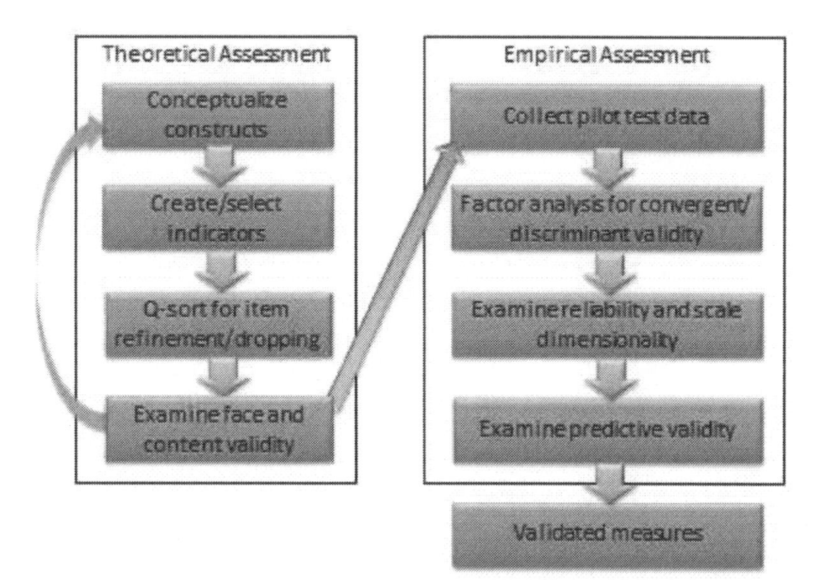

Figure 7.4. An integrated approach to measurement validation

The integrated approach starts in the theoretical realm. The first step is conceptualizing the constructs of interest. This includes defining each construct and identifying their constituent domains and/or dimensions. Next, we select (or create) items or indicators for each construct based on our conceptualization of these construct, as described in the scaling procedure in Chapter 5. A literature review may also be helpful in indicator selection. Each item is reworded in a uniform manner using simple and easy-to-understand text. Following this step, a panel of expert judges (academics experienced in research methods and/or a representative set of target respondents) can be employed to examine each indicator and conduct a Q-sort analysis. In this analysis, each judge is given a list of all constructs with their conceptual definitions and a stack of index cards listing each indicator for each of the construct measures (one indicator per index card). Judges are then asked to independently read each index card, examine the clarity, readability, and semantic meaning of that item, and sort it with the construct where it seems to make the most sense, based on the construct definitions provided. Inter-rater reliability is assessed to examine the extent to which judges agreed with their classifications. Ambiguous items that were consistently missed by many judges may be reexamined, reworded, or dropped. The best items (say 10-15) for each construct are selected for further analysis. Each of the selected items is reexamined by judges for face validity and content validity. If an adequate set of items is not achieved at this stage, new items may have to be created based on the conceptual definition of the intended construct. Two or three rounds of Q-sort may be needed to arrive at reasonable agreement between judges on a set of items that best represents the constructs of interest.

Next, the validation procedure moves to the empirical realm. A research instrument is created comprising all of the refined construct items, and is administered to a pilot test group of representative respondents from the target population. Data collected is tabulated and subjected to correlational analysis or exploratory factor analysis using a software program such as SAS or SPSS for assessment of convergent and discriminant validity. Items that do not meet the expected norms of factor loading (same-factor loadings higher than 0.60, and cross-factor loadings less than 0.30) should be dropped at this stage. The remaining scales are evaluated for reliability using a measure of internal consistency such as Cronbach alpha. Scale dimensionality may also be verified at this stage, depending on whether the targeted constructs were conceptualized as being unidimensional or multi-dimensional. Next, evaluate the predictive ability of each construct within a theoretically specified nomological network of construct using regression analysis or structural equation modeling. If the construct measures satisfy most or all of the requirements of reliability and validity described in this chapter, we can be assured that our operationalized measures are reasonably adequate and accurate.

The integrated approach to measurement validation discussed here is quite demanding of researcher time and effort. Nonetheless, this elaborate multi-stage process is needed to ensure that measurement scales used in our research meets the expected norms of scientific research. Because inferences drawn using flawed or compromised scales are meaningless, scale validation and measurement remains one of the most important and involved phase of empirical research.

Chapter 8

Sampling

Sampling is the statistical process of selecting a subset (called a "sample") of a population of interest for purposes of making observations and statistical inferences about that population. Social science research is generally about inferring patterns of behaviors within specific populations. We cannot study entire populations because of feasibility and cost constraints, and hence, we must select a representative sample from the population of interest for observation and analysis. It is extremely important to choose a sample that is truly representative of the population so that the inferences derived from the sample can be generalized back to the population of interest. Improper and biased sampling is the primary reason for often divergent and erroneous inferences reported in opinion polls and exit polls conducted by different polling groups such as CNN/Gallup Poll, ABC, and CBS, prior to every U.S. Presidential elections.

The Sampling Process

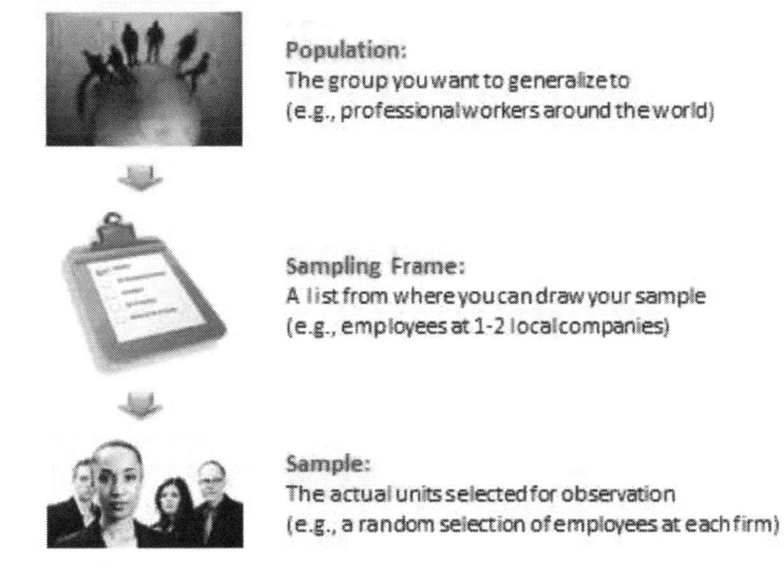

Figure 8.1. The sampling process

The sampling process comprises of several stage. The first stage is defining the target population. A **population** can be defined as all people or items (**unit of analysis**) with the characteristics that one wishes to study. The unit of analysis may be a person, group, organization, country, object, or any other entity that you wish to draw scientific inferences about. Sometimes the population is obvious. For example, if a manufacturer wants to determine whether finished goods manufactured at a production line meets certain quality requirements or must be scrapped and reworked, then the population consists of the entire set of finished goods manufactured at that production facility. At other times, the target population may be a little harder to understand. If you wish to identify the primary drivers of academic learning among high school students, then what is your target population: high school students, their teachers, school principals, or parents? The right answer in this case is high school students, because you are interested in their performance, not the performance of their teachers, parents, or schools. Likewise, if you wish to analyze the behavior of roulette wheels to identify biased wheels, your population of interest is not different observations from a single roulette wheel, but different roulette wheels (i.e., their behavior over an infinite set of wheels).

The second step in the sampling process is to choose a **sampling frame**. This is an accessible section of the target population (usually a list with contact information) from where a sample can be drawn. If your target population is professional employees at work, because you cannot access all professional employees around the world, a more realistic sampling frame will be employee lists of one or two local companies that are willing to participate in your study. If your target population is organizations, then the Fortune 500 list of firms or the Standard & Poor's (S&P) list of firms registered with the New York Stock exchange may be acceptable sampling frames.

Note that sampling frames may not entirely be representative of the population at large, and if so, inferences derived by such a sample may not be generalizable to the population. For instance, if your target population is organizational employees at large (e.g., you wish to study employee self-esteem in this population) and your sampling frame is employees at automotive companies in the American Midwest, findings from such groups may not even be generalizable to the American workforce at large, let alone the global workplace. This is because the American auto industry has been under severe competitive pressures for the last 50 years and has seen numerous episodes of reorganization and downsizing, possibly resulting in low employee morale and self-esteem. Furthermore, the majority of the American workforce is employed in service industries or in small businesses, and not in automotive industry. Hence, a sample of American auto industry employees is not particularly representative of the American workforce. Likewise, the Fortune 500 list includes the 500 largest American enterprises, which is not representative of all American firms in general, most of which are medium and small-sized firms rather than large firms, and is therefore, a biased sampling frame. In contrast, the S&P list will allow you to select large, medium, and/or small companies, depending on whether you use the S&P large-cap, mid-cap, or small-cap lists, but includes publicly traded firms (and not private firms) and hence still biased. Also note that the population from which a sample is drawn may not necessarily be the same as the population about which we actually want information. For example, if a researcher wants to the success rate of a new "quit smoking" program, then the target population is the universe of smokers who had access to this program, which may be an unknown population. Hence, the researcher may sample patients arriving at a local medical facility for smoking cessation treatment, some of whom may not have had exposure to this particular "quit smoking" program, in which case, the sampling frame does not correspond to the population of interest.

The last step in sampling is choosing a sample from the sampling frame using a well-defined sampling technique. Sampling techniques can be grouped into two broad categories: probability (random) sampling and non-probability sampling. Probability sampling is ideal if generalizability of results is important for your study, but there may be unique circumstances where non-probability sampling can also be justified. These techniques are discussed in the next two sections.

Probability Sampling

Probability sampling is a technique in which every unit in the population has a chance (non-zero probability) of being selected in the sample, and this chance can be accurately determined. Sample statistics thus produced, such as sample mean or standard deviation, are unbiased estimates of population parameters, as long as the sampled units are weighted according to their probability of selection. All probability sampling have two attributes in common: (1) every unit in the population has a known non-zero probability of being sampled, and (2) the sampling procedure involves random selection at some point. The different types of probability sampling techniques include:

Simple random sampling. In this technique, all possible subsets of a population (more accurately, of a sampling frame) are given an equal probability of being selected. The probability of selecting any set of n units out of a total of N units in a sampling frame is ${}^{N}C_{n}$. Hence, sample statistics are unbiased estimates of population parameters, without any weighting. Simple random sampling involves randomly selecting respondents from a sampling frame, but with large sampling frames, usually a table of random numbers or a computerized random number generator is used. For instance, if you wish to select 200 firms to survey from a list of 1000 firms, if this list is entered into a spreadsheet like Excel, you can use Excel's RAND() function to generate random numbers for each of the 1000 clients on that list. Next, you sort the list in increasing order of their corresponding random number, and select the first 200 clients on that sorted list. This is the simplest of all probability sampling techniques; however, the simplicity is also the strength of this technique. Because the sampling frame is not subdivided or partitioned, the sample is unbiased and the inferences are most generalizable amongst all probability sampling techniques.

Systematic sampling. In this technique, the sampling frame is ordered according to some criteria and elements are selected at regular intervals through that ordered list. Systematic sampling involves a random start and then proceeds with the selection of every k^{th} element from that point onwards, where $k = N/n$, where k is the ratio of sampling frame size N and the desired sample size n, and is formally called the *sampling ratio*. It is important that the starting point is not automatically the first in the list, but is instead randomly chosen from within the first k elements on the list. In our previous example of selecting 200 firms from a list of 1000 firms, you can sort the 1000 firms in increasing (or decreasing) order of their size (i.e., employee count or annual revenues), randomly select one of the first five firms on the sorted list, and then select every fifth firm on the list. This process will ensure that there is no overrepresentation of large or small firms in your sample, but rather that firms of all sizes are generally uniformly represented, as it is in your sampling frame. In other words, the sample is representative of the population, at least on the basis of the sorting criterion.

Stratified sampling. In stratified sampling, the sampling frame is divided into homogeneous and non-overlapping subgroups (called "strata"), and a simple random sample is drawn within each subgroup. In the previous example of selecting 200 firms from a list of 1000 firms, you can start by categorizing the firms based on their size as large (more than 500 employees), medium (between 50 and 500 employees), and small (less than 50 employees). You can then randomly select 67 firms from each subgroup to make up your sample of 200 firms. However, since there are many more small firms in a sampling frame than large firms, having an equal number of small, medium, and large firms will make the sample less representative of the population (i.e., biased in favor of large firms that are fewer in number in the target population). This is called *non-proportional* stratified sampling because the proportion of sample within each subgroup does not reflect the proportions in the sampling frame (or the population of interest), and the smaller subgroup (large-sized firms) is *over-sampled*. An alternative technique will be to select subgroup samples in proportion to their size in the population. For instance, if there are 100 large firms, 300 mid-sized firms, and 600 small firms, you can sample 20 firms from the "large" group, 60 from the "medium" group and 120 from the "small" group. In this case, the proportional distribution of firms in the population is retained in the sample, and hence this technique is called *proportional* stratified sampling. Note that the non-proportional approach is particularly effective in representing small subgroups, such as large-sized firms, and is not necessarily less representative of the population compared to the proportional approach, as long as the findings of the non-proportional approach is weighted in accordance to a subgroup's proportion in the overall population.

Cluster sampling. If you have a population dispersed over a wide geographic region, it may not be feasible to conduct a simple random sampling of the entire population. In such case, it may be reasonable to divide the population into "clusters" (usually along geographic boundaries), randomly sample a few clusters, and measure *all* units within that cluster. For instance, if you wish to sample city governments in the state of New York, rather than travel all over the state to interview key city officials (as you may have to do with a simple random sample), you can cluster these governments based on their counties, randomly select a set of three counties, and then interview officials from *every* official in those counties. However, depending on between-cluster differences, the variability of sample estimates in a cluster sample will generally be higher than that of a simple random sample, and hence the results are less generalizable to the population than those obtained from simple random samples.

Matched-pairs sampling. Sometimes, researchers may want to compare two subgroups within one population based on a specific criterion. For instance, why are some firms consistently more profitable than other firms? To conduct such a study, you would have to categorize a sampling frame of firms into "high profitable" firms and "low profitable firms" based on gross margins, earnings per share, or some other measure of profitability. You would then select a simple random sample of firms in one subgroup, and match each firm in this group with a firm in the second subgroup, based on its size, industry segment, and/or other matching criteria. Now, you have two matched samples of high-profitability and low-profitability firms that you can study in greater detail. Such matched-pairs sampling technique is often an ideal way of understanding bipolar differences between different subgroups within a given population.

Multi-stage sampling. The probability sampling techniques described previously are all examples of single-stage sampling techniques. Depending on your sampling needs, you may combine these single-stage techniques to conduct multi-stage sampling. For instance, you can stratify a list of businesses based on firm size, and then conduct systematic sampling within each stratum. This is a two-stage combination of stratified and systematic sampling. Likewise, you can start with a cluster of school districts in the state of New York, and within each cluster, select a simple random sample of schools; within each school, select a simple random sample of grade levels; and within each grade level, select a simple random sample of students for study. In this case, you have a four-stage sampling process consisting of cluster and simple random sampling.

Non-Probability Sampling

Nonprobability sampling is a sampling technique in which some units of the population have *zero* chance of selection or where the probability of selection cannot be accurately determined. Typically, units are selected based on certain non-random criteria, such as quota or convenience. Because selection is non-random, nonprobability sampling does not allow the estimation of sampling errors, and may be subjected to a sampling bias. Therefore, information from a sample cannot be generalized back to the population. Types of non-probability sampling techniques include:

Convenience sampling. Also called accidental or opportunity sampling, this is a technique in which a sample is drawn from that part of the population that is close to hand, readily available, or convenient. For instance, if you stand outside a shopping center and hand out questionnaire surveys to people or interview them as they walk in, the sample of respondents you will obtain will be a convenience sample. This is a non-probability sample because you are systematically excluding all people who shop at other shopping centers. The opinions that you would get from your chosen sample may reflect the unique characteristics of this shopping center such as the nature of its stores (e.g., high end-stores will attract a more affluent demographic), the demographic profile of its patrons, or its location (e.g., a shopping center close to a university will attract primarily university students with unique purchase habits), and therefore may not be representative of the opinions of the shopper population at large. Hence, the scientific generalizability of such observations will be very limited. Other examples of convenience sampling are sampling students registered in a certain class or sampling patients arriving at a certain medical clinic. This type of sampling is most useful for pilot testing, where the goal is instrument testing or measurement validation rather than obtaining generalizable inferences.

Quota sampling. In this technique, the population is segmented into mutually-exclusive subgroups (just as in stratified sampling), and then a non-random set of observations is chosen from each subgroup to meet a predefined quota. In **proportional quota sampling**, the proportion of respondents in each subgroup should match that of the population. For instance, if the American population consists of 70% Caucasians, 15% Hispanic-Americans, and 13% African-Americans, and you wish to understand their voting preferences in an sample of 98 people, you can stand outside a shopping center and ask people their voting preferences. But you will have to stop asking Hispanic-looking people when you have 15 responses from that subgroup (or African-Americans when you have 13 responses) even as you continue sampling other ethnic groups, so that the ethnic composition of your sample matches that of the general American population. **Non-proportional quota sampling** is less restrictive in that you don't have to achieve a proportional representation, but perhaps meet a minimum size in each subgroup. In this case, you may decide to have 50 respondents from each of the three ethnic subgroups (Caucasians, Hispanic-Americans, and African-Americans), and stop when your quota for each subgroup is reached. Neither type of quota sampling will be representative of the American population, since depending on whether your study was conducted in a shopping center in New York or Kansas, your results may be entirely different. The non-proportional technique is even less representative of the population but may be useful in that it allows capturing the opinions of small and underrepresented groups through oversampling.

Expert sampling. This is a technique where respondents are chosen in a non-random manner based on their expertise on the phenomenon being studied. For instance, in order to understand the impacts of a new governmental policy such as the Sarbanes-Oxley Act, you can sample an group of corporate accountants who are familiar with this act. The advantage of this approach is that since experts tend to be more familiar with the subject matter than non-experts, opinions from a sample of experts are more credible than a sample that includes both experts and non-experts, although the findings are still not generalizable to the overall population at large.

Snowball sampling. In snowball sampling, you start by identifying a few respondents that match the criteria for inclusion in your study, and then ask them to recommend others they know who also meet your selection criteria. For instance, if you wish to survey computer network administrators and you know of only one or two such people, you can start with them and ask them to recommend others who also do network administration. Although this method hardly leads to representative samples, it may sometimes be the only way to reach hard-to-reach populations or when no sampling frame is available.

Statistics of Sampling

In the preceding sections, we introduced terms such as population parameter, sample statistic, and sampling bias. In this section, we will try to understand what these terms mean and how they are related to each other.

When you measure a certain observation from a given unit, such as a person's response to a Likert-scaled item, that observation is called a response (see Figure 8.2). In other words, a **response** is a measurement value provided by a sampled unit. Each respondent will give you different responses to different items in an instrument. Responses from different respondents to the same item or observation can be graphed into a **frequency distribution** based on their frequency of occurrences. For a large number of responses in a sample, this frequency distribution tends to resemble a bell-shaped curve called a **normal distribution**, which can be used to estimate overall characteristics of the entire sample, such as sample mean (average of all observations in a sample) or standard deviation (variability or spread of observations in a sample). These sample estimates are called **sample statistics** (a "statistic" is a value that is estimated from observed data). Populations also have means and standard deviations that could be obtained if we could sample the entire population. However, since the entire population can never be sampled, population characteristics are always unknown, and are called **population parameters** (and not "statistic" because they are not statistically estimated from data). Sample statistics may differ from population parameters if the sample is not perfectly representative of the population; the difference between the two is called **sampling error**. Theoretically, if we could gradually increase the sample size so that the sample approaches closer and closer to the population, then sampling error will decrease and a sample statistic will increasingly approximate the corresponding population parameter.

If a sample is truly representative of the population, then the *estimated* sample statistics should be identical to corresponding *theoretical* population parameters. How do we know if the sample statistics are at least reasonably close to the population parameters? Here, we need to understand the concept of a **sampling distribution**. Imagine that you took three different random samples from a given population, as shown in Figure 8.3, and for each sample, you derived sample statistics such as sample mean and standard deviation. If each random sample was truly representative of the population, then your three sample means from the three random samples will be identical (and equal to the population parameter), and the variability in sample means will be zero. But this is extremely unlikely, given that each random sample will likely constitute a different subset of the population, and hence, their means may be slightly different from each other. However, you can take these three sample means and plot a frequency histogram of sample means. If the number of such samples increases from three to 10 to 100, the frequency histogram becomes a sampling distribution. Hence, a sampling distribution is a frequency distribution of a *sample statistic* (like sample mean) from a *set of samples*, while the commonly referenced frequency distribution is the distribution of a *response* (observation) from *a single sample*. Just like a frequency distribution, the sampling distribution will also tend to have more sample statistics clustered around the mean (which presumably is an estimate of a population parameter), with fewer values scattered around the mean. With an infinitely large number of samples, this distribution will approach a normal distribution. The variability or spread of a sample statistic in a sampling distribution (i.e., the standard deviation of a sampling statistic) is called its **standard error**. In contrast, the term standard deviation is reserved for variability of an observed response from a single sample.

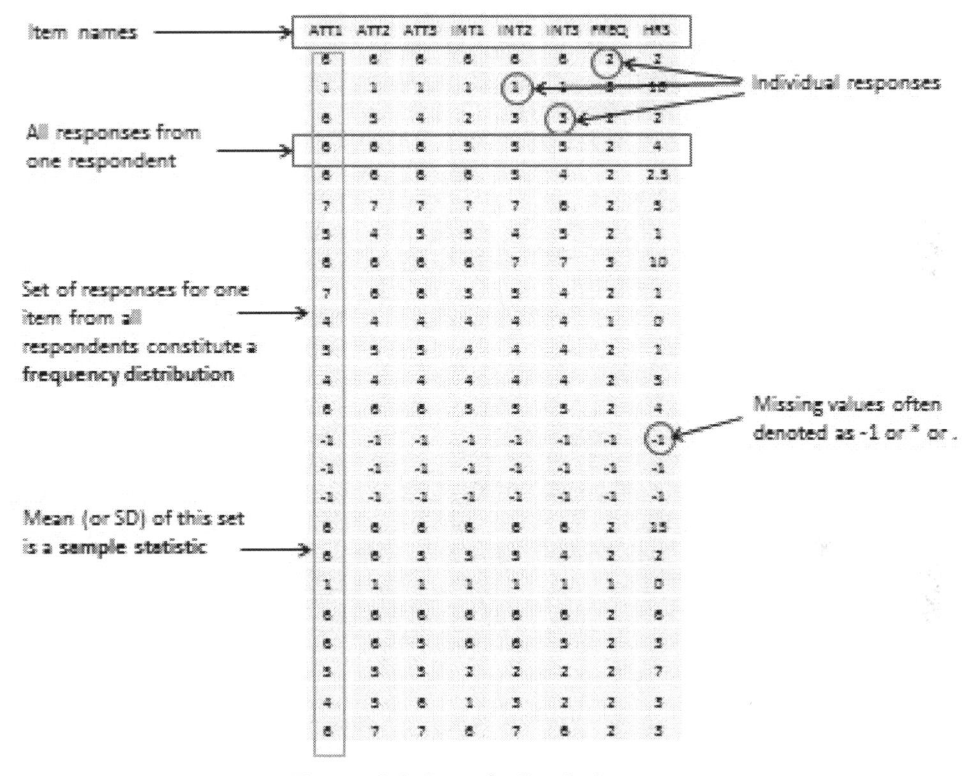

Figure 8.2. Sample Statistic

The mean value of a sample statistic in a sampling distribution is presumed to be an estimate of the unknown population parameter. Based on the spread of this sampling distribution (i.e., based on standard error), it is also possible to estimate confidence intervals for that prediction population parameter. **Confidence interval** is the estimated probability that a population parameter lies within a specific interval of sample statistic values. All normal distributions tend to follow a 68-95-99 percent rule (see Figure 8.4), which says that over 68% of the cases in the distribution lie within one standard deviation of the mean value ($\mu \pm 1\sigma$), over 95% of the cases in the distribution lie within two standard deviations of the mean ($\mu \pm 2\sigma$), and over 99% of the cases in the distribution lie within three standard deviations of the mean value ($\mu \pm 3\sigma$). Since a sampling distribution with an infinite number of samples will approach a normal distribution, the same 68-95-99 rule applies, and it can be said that:

- (Sample statistic $\pm$ one standard error) represents a 68% confidence interval for the population parameter.

- (Sample statistic ± two standard errors) represents a 95% confidence interval for the population parameter.
- (Sample statistic ± three standard errors) represents a 99% confidence interval for the population parameter.

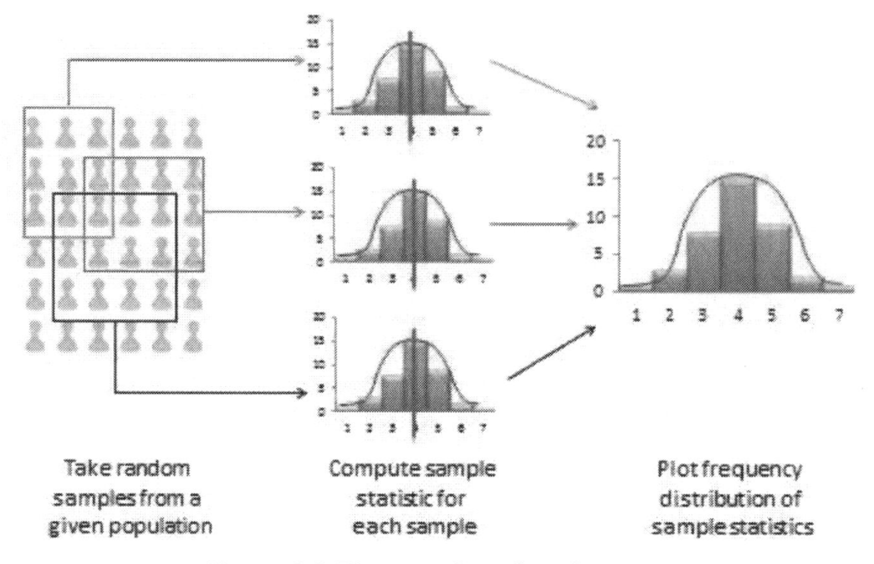

Figure 8.3. The sampling distribution

A sample is "biased" (i.e., not representative of the population) if its sampling distribution cannot be estimated or if the sampling distribution violates the 68-95-99 percent rule. As an aside, note that in most regression analysis where we examine the significance of regression coefficients with p<0.05, we are attempting to see if the sampling statistic (regression coefficient) predicts the corresponding population parameter (true effect size) with a 95% confidence interval. Interestingly, the "six sigma" standard attempts to identify manufacturing defects outside the 99% confidence interval or six standard deviations (standard deviation is represented using the Greek letter sigma), representing significance testing at p<0.01.

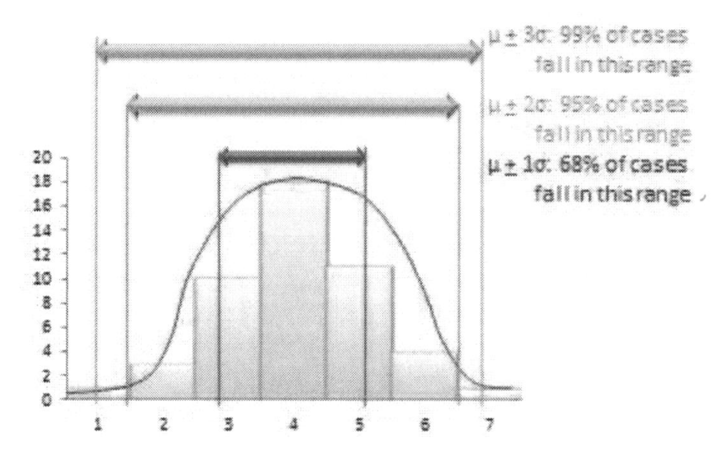

Figure 8.4. The 68-95-99 percent rule for confidence interval

Chapter 9

Survey Research

Survey research a research method involving the use of standardized questionnaires or interviews to collect data about people and their preferences, thoughts, and behaviors in a systematic manner. Although census surveys were conducted as early as Ancient Egypt, survey as a formal research method was pioneered in the 1930-40s by sociologist Paul Lazarsfeld to examine the effects of the radio on political opinion formation of the United States. This method has since become a very popular method for quantitative research in the social sciences.

The survey method can be used for descriptive, exploratory, or explanatory research. This method is best suited for studies that have individual people as the unit of analysis. Although other units of analysis, such as groups, organizations or dyads (pairs of organizations, such as buyers and sellers), are also studied using surveys, such studies often use a specific person from each unit as a "key informant" or a "proxy" for that unit, and such surveys may be subject to respondent bias if the informant chosen does not have adequate knowledge or has a biased opinion about the phenomenon of interest. For instance, Chief Executive Officers may not adequately know employee's perceptions or teamwork in their own companies, and may therefore be the wrong informant for studies of team dynamics or employee self-esteem.

Survey research has several inherent strengths compared to other research methods. First, surveys are an excellent vehicle for measuring a wide variety of unobservable data, such as people's preferences (e.g., political orientation), traits (e.g., self-esteem), attitudes (e.g., toward immigrants), beliefs (e.g., about a new law), behaviors (e.g., smoking or drinking behavior), or factual information (e.g., income). Second, survey research is also ideally suited for remotely collecting data about a population that is too large to observe directly. A large area, such as an entire country, can be covered using mail-in, electronic mail, or telephone surveys using meticulous sampling to ensure that the population is adequately represented in a small sample. Third, due to their unobtrusive nature and the ability to respond at one's convenience, questionnaire surveys are preferred by some respondents. Fourth, interviews may be the only way of reaching certain population groups such as the homeless or illegal immigrants for which there is no sampling frame available. Fifth, large sample surveys may allow detection of small effects even while analyzing multiple variables, and depending on the survey design, may also allow comparative analysis of population subgroups (i.e., within-group and between-group analysis). Sixth, survey research is economical in terms of researcher time, effort and cost than most other methods such as experimental research and case research. At the same time, survey research also has some unique disadvantages. It is subject to a large number of biases such as non-response bias, sampling bias, social desirability bias, and recall bias, as discussed in the last section of this chapter.

Depending on how the data is collected, survey research can be divided into two broad categories: questionnaire surveys (which may be mail-in, group-administered, or online surveys), and interview surveys (which may be personal, telephone, or focus group interviews). Questionnaires are instruments that are completed in writing by respondents, while interviews are completed by the interviewer based on verbal responses provided by respondents. As discussed below, each type has its own strengths and weaknesses, in terms of their costs, coverage of the target population, and researcher's flexibility in asking questions.

Questionnaire Surveys

Invented by Sir Francis Galton, a **questionnaire** is a research instrument consisting of a set of questions (items) intended to capture responses from respondents in a standardized manner. Questions may be unstructured or structured. Unstructured questions ask respondents to provide a response in their own words, while structured questions ask respondents to select an answer from a given set of choices. Subjects' responses to individual questions (items) on a structured questionnaire may be aggregated into a composite scale or index for statistical analysis. Questions should be designed such that respondents are able to read, understand, and respond to them in a meaningful way, and hence the survey method may not be appropriate or practical for certain demographic groups such as children or the illiterate.

Most questionnaire surveys tend to be **self-administered mail surveys**, where the same questionnaire is mailed to a large number of people, and willing respondents can complete the survey at their convenience and return it in postage-prepaid envelopes. Mail surveys are advantageous in that they are unobtrusive, and they are inexpensive to administer, since bulk postage is cheap in most countries. However, response rates from mail surveys tend to be quite low since most people tend to ignore survey requests. There may also be long delays (several months) in respondents' completing and returning the survey (or they may simply lose it). Hence, the researcher must continuously monitor responses as they are being returned, track and send reminders to non-respondents repeated reminders (two or three reminders at intervals of one to 1.5 months is ideal). Questionnaire surveys are also not well-suited for issues that require clarification on the part of the respondent or those that require detailed written responses. Longitudinal designs can be used to survey the same set of respondents at different times, but response rates tend to fall precipitously from one survey to the next.

A second type of survey is **group-administered questionnaire**. A sample of respondents is brought together at a common place and time, and each respondent is asked to complete the survey questionnaire while in that room. Respondents enter their responses independently without interacting with each other. This format is convenient for the researcher, and high response rate is assured. If respondents do not understand any specific question, they can ask for clarification. In many organizations, it is relatively easy to assemble a group of employees in a conference room or lunch room, especially if the survey is approved by corporate executives.

A more recent type of questionnaire survey is an online or web survey. These surveys are administered over the Internet using interactive forms. Respondents may receive an electronic mail request for participation in the survey with a link to an online website where the survey may be completed. Alternatively, the survey may be embedded into an e-mail, and can be completed and returned via e-mail. These surveys are very inexpensive to administer, results are instantly recorded in an online database, and the survey can be easily modified if needed. However, if the survey website is not password-protected or designed to prevent multiple submissions, the responses can be easily compromised. Furthermore, sampling bias may be a significant issue since the survey cannot reach people that do not have computer or Internet access, such as many of the poor, senior, and minority groups, and the respondent sample is skewed toward an younger demographic who are online much of the time and have the time and ability to complete such surveys. Computing the response rate may be problematic, if the survey link is posted on listservs or bulletin boards instead of being e-mailed directly to targeted respondents. For these reasons, many researchers prefer dual-media surveys (e.g., mail survey and online survey), allowing respondents to select their preferred method of response.

Constructing a survey questionnaire is an art. Numerous decisions must be made about the content of questions, their wording, format, and sequencing, all of which can have important consequences for the survey responses.

Response formats. Survey questions may be structured or unstructured. Responses to structured questions are captured using one of the following response formats:

- *Dichotomous response*, where respondents are asked to select one of two possible choices, such as true/false, yes/no, or agree/disagree. An example of such a question is: Do you think that the death penalty is justified under some circumstances (circle one): yes / no.
- *Nominal response*, where respondents are presented with more than two unordered options, such as: What is your industry of employment: manufacturing / consumer services / retail / education / healthcare / tourism & hospitality / other.
- *Ordinal response*, where respondents have more than two ordered options, such as: what is your highest level of education: high school / college degree / graduate studies.
- *Interval-level response*, where respondents are presented with a 5-point or 7-point Likert scale, semantic differential scale, or Guttman scale. Each of these scale types were discussed in a previous chapter.
- *Continuous response*, where respondents enter a continuous (ratio-scaled) value with a meaningful zero point, such as their age or tenure in a firm. These responses generally tend to be of the fill-in-the blanks type.

Question content and wording. Responses obtained in survey research are very sensitive to the types of questions asked. Poorly framed or ambiguous questions will likely result in meaningless responses with very little value. Dillman (1978) recommends several rules for creating good survey questions. Every single question in a survey should be carefully scrutinized for the following issues:

- *Is the question clear and understandable:* Survey questions should be stated in a very simple language, preferably in active voice, and without complicated words or jargon that may not be understood by a typical respondent. All questions in the questionnaire should be worded in a similar manner to make it easy for respondents to read and understand them. The only exception is if your survey is targeted at a specialized group of respondents, such as doctors, lawyers and researchers, who use such jargon in their everyday environment.

- *Is the question worded in a negative manner:* Negatively worded questions, such as should your local government not raise taxes, tend to confuse many responses and lead to inaccurate responses. Such questions should be avoided, and in all cases, avoid double-negatives.
- *Is the question ambiguous:* Survey questions should not words or expressions that may be interpreted differently by different respondents (e.g., words like "any" or "just"). For instance, if you ask a respondent, what is your annual income, it is unclear whether you referring to salary/wages, or also dividend, rental, and other income, whether you referring to personal income, family income (including spouse's wages), or personal and business income? Different interpretation by different respondents will lead to incomparable responses that cannot be interpreted correctly.
- *Does the question have biased or value-laden words:* Bias refers to any property of a question that encourages subjects to answer in a certain way. Kenneth Rasinky (1989) examined several studies on people's attitude toward government spending, and observed that respondents tend to indicate stronger support for "assistance to the poor" and less for "welfare", even though both terms had the same meaning. In this study, more support was also observed for "halting rising crime rate" (and less for "law enforcement"), "solving problems of big cities" (and less for "assistance to big cities"), and "dealing with drug addiction" (and less for "drug rehabilitation"). A biased language or tone tends to skew observed responses. It is often difficult to anticipate in advance the biasing wording, but to the greatest extent possible, survey questions should be carefully scrutinized to avoid biased language.
- *Is the question double-barreled:* Double-barreled questions are those that can have multiple answers. For example, are you satisfied with the hardware and software provided for your work? In this example, how should a respondent answer if he/she is satisfied with the hardware but not with the software or vice versa? It is always advisable to separate double-barreled questions into separate questions: (1) are you satisfied with the hardware provided for your work, and (2) are you satisfied with the software provided for your work. Another example: does your family favor public television? Some people may favor public TV for themselves, but favor certain cable TV programs such as *Sesame Street* for their children.
- *Is the question too general:* Sometimes, questions that are too general may not accurately convey respondents' perceptions. If you asked someone how they liked a certain book and provide a response scale ranging from "not at all" to "extremely well", if that person selected "extremely well", what does he/she mean? Instead, ask more specific behavioral questions, such as will you recommend this book to others, or do you plan to read other books by the same author? Likewise, instead of asking how big is your firm (which may be interpreted differently by respondents), ask how many people work for your firm, and/or what is the annual revenues of your firm, which are both measures of firm size.
- *Is the question too detailed:* Avoid unnecessarily detailed questions that serve no specific research purpose. For instance, do you need the age of each child in a household or is just the number of children in the household acceptable? However, if unsure, it is better to err on the side of details than generality.
- *Is the question presumptuous:* If you ask, what do you see are the benefits of a tax cut, you are presuming that the respondent sees the tax cut as beneficial. But many people may not view tax cuts as being beneficial, because tax cuts generally lead to lesser

funding for public schools, larger class sizes, and fewer public services such as police, ambulance, and fire service. Avoid questions with built-in presumptions.

- *Is the question imaginary:* A popular question in many television game shows is "if you won a million dollars on this show, how will you plan to spend it?" Most respondents have never been faced with such an amount of money and have never thought about it (most don't even know that after taxes, they will get only about $640,000 or so in the United States, and in many cases, that amount is spread over a 20-year period, so that their net present value is even less), and so their answers tend to be quite random, such as take a tour around the world, buy a restaurant or bar, spend on education, save for retirement, help parents or children, or have a lavish wedding. Imaginary questions have imaginary answers, which cannot be used for making scientific inferences.
- *Do respondents have the information needed to correctly answer the question:* Often times, we assume that subjects have the necessary information to answer a question, when in reality, they do not. Even if a response is obtained, in such case, the responses tend to be inaccurate, given their lack of knowledge about the question being asked. For instance, we should not ask the CEO of a company about day-to-day operational details that they may not be aware of, or asking teachers about how much their students are learning, or asking high-schoolers "Do you think the US Government acted appropriately in the Bay of Pigs crisis?"

Question sequencing. In general, questions should flow logically from one to the next. To achieve the best response rates, questions should flow from the least sensitive to the most sensitive, from the factual and behavioral to the attitudinal, and from the more general to the more specific. Some general rules for question sequencing:

- Start with easy non-threatening questions that can be easily recalled. Good options are demographics (age, gender, education level) for individual-level surveys and firmographics (employee count, annual revenues, industry) for firm-level surveys.
- Never start with an open ended question.
- If following an historical sequence of events, follow a chronological order from earliest to latest.
- Ask about one topic at a time. When switching topics, use a transition, such as "The next section examines your opinions about ..."
- Use filter or contingency questions as needed, such as: "If you answered "yes" to question 5, please proceed to Section 2. If you answered "no" go to Section 3."

Other golden rules. Do unto your respondents what you would have them do unto you. Be attentive and appreciative of respondents' time, attention, trust, and confidentiality of personal information. Always practice the following strategies for all survey research:

- People's time is valuable. Be respectful of their time. Keep your survey as short as possible and limit it to what is absolutely necessary. Respondents do not like spending more than 10-15 minutes on any survey, no matter how important it is. Longer surveys tend to dramatically lower response rates.
- Always assure respondents about the confidentiality of their responses, and how you will use their data (e.g., for academic research) and how the results will be reported (usually, in the aggregate).
- For organizational surveys, assure respondents that you will send them a copy of the final results, and make sure that you follow up with your promise.
- Thank your respondents for their participation in your study.

- Finally, always pretest your questionnaire, at least using a convenience sample, before administering it to respondents in a field setting. Such pretesting may uncover ambiguity, lack of clarity, or biases in question wording, which should be eliminated before administering to the intended sample.

Interview Survey

Interviews are a more personalized form of data collection method than questionnaires, and are conducted by trained interviewers using the same research protocol as questionnaire surveys (i.e., a standardized set of questions). However, unlike a questionnaire, the interview script may contain special instructions for the interviewer that is not seen by respondents, and may include space for the interviewer to record personal observations and comments. In addition, unlike mail surveys, the interviewer has the opportunity to clarify any issues raised by the respondent or ask probing or follow-up questions. However, interviews are time-consuming and resource-intensive. Special interviewing skills are needed on part of the interviewer. The interviewer is also considered to be part of the measurement instrument, and must proactively strive not to artificially bias the observed responses.

The most typical form of interview is personal or **face-to-face interview**, where the interviewer works directly with the respondent to ask questions and record their responses. Personal interviews may be conducted at the respondent's home or office location. This approach may even be favored by some respondents, while others may feel uncomfortable in allowing a stranger in their homes. However, skilled interviewers can persuade respondents to cooperate, dramatically improving response rates.

A variation of the personal interview is a group interview, also called **focus group**. In this technique, a small group of respondents (usually 6-10 respondents) are interviewed together in a common location. The interviewer is essentially a facilitator whose job is to lead the discussion, and ensure that every person has an opportunity to respond. Focus groups allow deeper examination of complex issues than other forms of survey research, because when people hear others talk, it often triggers responses or ideas that they did not think about before. However, focus group discussion may be dominated by a dominant personality, and some individuals may be reluctant to voice their opinions in front of their peers or superiors, especially while dealing with a sensitive issue such as employee underperformance or office politics. Because of their small sample size, focus groups are usually used for exploratory research rather than descriptive or explanatory research.

A third type of interview survey is **telephone interviews**. In this technique, interviewers contact potential respondents over the phone, typically based on a random selection of people from a telephone directory, to ask a standard set of survey questions. A more recent and technologically advanced approach is computer-assisted telephone interviewing (CATI), increasing being used by academic, government, and commercial survey researchers, where the interviewer is a telephone operator, who is guided through the interview process by a computer program displaying instructions and questions to be asked on a computer screen. The system also selects respondents randomly using a random digit dialing technique, and records responses using voice capture technology. Once respondents are on the phone, higher response rates can be obtained. This technique is not ideal for rural areas where telephone density is low, and also cannot be used for communicating non-audio information such as graphics or product demonstrations.

Role of interviewer. The interviewer has a complex and multi-faceted role in the interview process, which includes the following tasks:

- *Prepare for the interview:* Since the interviewer is in the forefront of the data collection effort, the quality of data collected depends heavily on how well the interviewer is trained to do the job. The interviewer must be trained in the interview process and the survey method, and also be familiar with the purpose of the study, how responses will be stored and used, and sources of interviewer bias. He/she should also rehearse and time the interview prior to the formal study.
- *Locate and enlist the cooperation of respondents:* Particularly in personal, in-home surveys, the interviewer must locate specific addresses, and work around respondents' schedule sometimes at undesirable times such as during weekends. They should also be like a salesperson, selling the idea of participating in the study.
- *Motivate respondents:* Respondents often feed off the motivation of the interviewer. If the interviewer is disinterested or inattentive, respondents won't be motivated to provide useful or informative responses either. The interviewer must demonstrate enthusiasm about the study, communicate the importance of the research to respondents, and be attentive to respondents' needs throughout the interview.
- *Clarify any confusion or concerns:* Interviewers must be able to think on their feet and address unanticipated concerns or objections raised by respondents to the respondents' satisfaction. Additionally, they should ask probing questions as necessary even if such questions are not in the script.
- *Observe quality of response:* The interviewer is in the best position to judge the quality of information collected, and may supplement responses obtained using personal observations of gestures or body language as appropriate.

Conducting the interview. Before the interview, the interviewer should prepare a kit to carry to the interview session, consisting of a cover letter from the principal investigator or sponsor, adequate copies of the survey instrument, photo identification, and a telephone number for respondents to call to verify the interviewer's authenticity. The interviewer should also try to call respondents ahead of time to set up an appointment if possible. To start the interview, he/she should speak in an imperative and confident tone, such as "I'd like to take a few minutes of your time to interview you for a very important study," instead of "May I come in to do an interview?" He/she should introduce himself/herself, present personal credentials, explain the purpose of the study in 1-2 sentences, and assure confidentiality of respondents' comments and voluntariness of their participation, all in less than a minute. No big words or jargon should be used, and no details should be provided unless specifically requested. If the interviewer wishes to tape-record the interview, he/she should ask for respondent's explicit permission before doing so. Even if the interview is recorded, the interview must take notes on key issues, probes, or verbatim phrases.

During the interview, the interviewer should follow the questionnaire script and ask questions exactly as written, and not change the words to make the question sound friendlier. They should also not change the order of questions or skip any question that may have been answered earlier. Any issues with the questions should be discussed during rehearsal prior to the actual interview sessions. The interviewer should not finish the respondent's sentences. If the respondent gives a brief cursory answer, the interviewer should probe the respondent to elicit a more thoughtful, thorough response. Some useful probing techniques are:

- *The silent probe:* Just pausing and waiting (without going into the next question) may suggest to respondents that the interviewer is waiting for more detailed response.
- *Overt encouragement:* Occasional "uh-huh" or "okay" may encourage the respondent to go into greater details. However, the interviewer must not express approval or disapproval of what was said by the respondent.
- *Ask for elaboration:* Such as "can you elaborate on that?" or "A minute ago, you were talking about an experience you had in high school. Can you tell me more about that?"
- *Reflection:* The interviewer can try the psychotherapist's trick of repeating what the respondent said. For instance, "What I'm hearing is that you found that experience very traumatic" and then pause and wait for the respondent to elaborate.

After the interview in completed, the interviewer should thank respondents for their time, tell them when to expect the results, and not leave hastily. Immediately after leaving, they should write down any notes or key observations that may help interpret the respondent's comments better.

Biases in Survey Research

Despite all of its strengths and advantages, survey research is often tainted with systematic biases that may invalidate some of the inferences derived from such surveys. Five such biases are the non-response bias, sampling bias, social desirability bias, recall bias, and common method bias.

Non-response bias. Survey research is generally notorious for its low response rates. A response rate of 15-20% is typical in a mail survey, even after two or three reminders. If the majority of the targeted respondents fail to respond to a survey, then a legitimate concern is whether non-respondents are not responding due to a systematic reason, which may raise questions about the validity of the study's results. For instance, dissatisfied customers tend to be more vocal about their experience than satisfied customers, and are therefore more likely to respond to questionnaire surveys or interview requests than satisfied customers. Hence, any respondent sample is likely to have a higher proportion of dissatisfied customers than the underlying population from which it is drawn. In this instance, not only will the results lack generalizability, but the observed outcomes may also be an artifact of the biased sample. Several strategies may be employed to improve response rates:

- *Advance notification:* A short letter sent in advance to the targeted respondents soliciting their participation in an upcoming survey can prepare them in advance and improve their propensity to respond. The letter should state the purpose and importance of the study, mode of data collection (e.g., via a phone call, a survey form in the mail, etc.), and appreciation for their cooperation. A variation of this technique may request the respondent to return a postage-paid postcard indicating whether or not they are willing to participate in the study.
- *Relevance of content:* If a survey examines issues of relevance or importance to respondents, then they are more likely to respond than to surveys that don't matter to them.

- *Respondent-friendly questionnaire:* Shorter survey questionnaires tend to elicit higher response rates than longer questionnaires. Furthermore, questions that are clear, non-offensive, and easy to respond tend to attract higher response rates.
- *Endorsement:* For organizational surveys, it helps to gain endorsement from a senior executive attesting to the importance of the study to the organization. Such endorsement can be in the form of a cover letter or a letter of introduction, which can improve the researcher's credibility in the eyes of the respondents.
- *Follow-up requests:* Multiple follow-up requests may coax some non-respondents to respond, even if their responses are late.
- *Interviewer training:* Response rates for interviews can be improved with skilled interviewers trained on how to request interviews, use computerized dialing techniques to identify potential respondents, and schedule callbacks for respondents who could not be reached.
- *Incentives*: Response rates, at least with certain populations, may increase with the use of incentives in the form of cash or gift cards, giveaways such as pens or stress balls, entry into a lottery, draw or contest, discount coupons, promise of contribution to charity, and so forth.
- *Non-monetary incentives:* Businesses, in particular, are more prone to respond to non-monetary incentives than financial incentives. An example of such a non-monetary incentive is a benchmarking report comparing the business's individual response against the aggregate of all responses to a survey.
- *Confidentiality and privacy:* Finally, assurances that respondents' private data or responses will not fall into the hands of any third party, may help improve response rates.

Sampling bias. Telephone surveys conducted by calling a random sample of publicly available telephone numbers will systematically exclude people with unlisted telephone numbers, mobile phone numbers, and people who are unable to answer the phone (for instance, they are at work) when the survey is being conducted, and will include a disproportionate number of respondents who have land-line telephone service with listed phone numbers and people who stay home during much of the day, such as the unemployed, the disabled, and the elderly. Likewise, online surveys tend to include a disproportionate number of students and younger people who are constantly on the Internet, and systematically exclude people with limited or no access to computers or the Internet, such as the poor and the elderly. Similarly, questionnaire surveys tend to exclude children and the illiterate, who are unable to read, understand, or meaningfully respond to the questionnaire. A different kind of sampling bias relate to sampling the wrong population, such as asking teachers (or parents) about academic learning of their students (or children), or asking CEOs about operational details in their company. Such biases make the respondent sample unrepresentative of the intended population and hurt generalizability claims about inferences drawn from the biased sample.

Social desirability bias. Many respondents tend to avoid negative opinions or embarrassing comments about themselves, their employers, family, or friends. With negative questions such as do you think that your project team is dysfunctional, is there a lot of office politics in your workplace, or have you ever illegally downloaded music files from the Internet, the researcher may not get truthful responses. This tendency among respondents to "spin the truth" in order to portray themselves in a socially desirable manner is called the "social desirability bias", which hurts the validity of response obtained from survey research. There is practically no way of overcoming the social desirability bias in a questionnaire survey, but in an interview setting, an astute interviewer may be able to spot inconsistent answers and ask probing questions or use personal observations to supplement respondents' comments.

Recall bias. Responses to survey questions often depend on subjects' motivation, memory, and ability to respond. Particularly when dealing with events that happened in the distant past, respondents may not adequately remember their own motivations or behaviors or perhaps their memory of such events may have evolved with time and no longer retrievable. For instance, if a respondent to asked to describe his/her utilization of computer technology one year ago or even memorable childhood events like birthdays, their response may not be accurate due to difficulties with recall. One possible way of overcoming the recall bias is by anchoring respondent's memory in specific events as they happened, rather than asking them to recall their perceptions and motivations from memory.

Common method bias. Common method bias refers to the amount of spurious covariance shared between independent and dependent variables that are measured at the same point in time, such as in a cross-sectional survey, using the same instrument, such as a questionnaire. In such cases, the phenomenon under investigation may not be adequately separated from measurement artifacts. Standard statistical tests are available to test for common method bias, such as Harmon's single-factor test (Podsakoff et al. 2003), Lindell and Whitney's (2001) market variable technique, and so forth. This bias can be potentially avoided if the independent and dependent variables are measured at different points in time, using a longitudinal survey design, of if these variables are measured using different methods, such as computerized recording of dependent variable versus questionnaire-based self-rating of independent variables.

Chapter 10

Experimental Research

Experimental research, often considered to be the "gold standard" in research designs, is one of the most rigorous of all research designs. In this design, one or more independent variables are manipulated by the researcher (as treatments), subjects are randomly assigned to different treatment levels (random assignment), and the results of the treatments on outcomes (dependent variables) are observed. The unique strength of experimental research is its **internal validity** (causality) due to its ability to link cause and effect through treatment manipulation, while controlling for the spurious effect of extraneous variable.

Experimental research is best suited for explanatory research (rather than for descriptive or exploratory research), where the goal of the study is to examine cause-effect relationships. It also works well for research that involves a relatively limited and well-defined set of independent variables that can either be manipulated or controlled. Experimental research can be conducted in laboratory or field settings. **Laboratory experiments**, conducted in laboratory (artificial) settings, tend to be high in internal validity, but this comes at the cost of low **external validity** (generalizability), because the artificial (laboratory) setting in which the study is conducted may not reflect the real world. **Field experiments**, conducted in field settings such as in a real organization, and high in both internal and external validity. But such experiments are relatively rare, because of the difficulties associated with manipulating treatments and controlling for extraneous effects in a field setting.

Experimental research can be grouped into two broad categories: true experimental designs and quasi-experimental designs. Both designs require treatment manipulation, but while true experiments also require random assignment, quasi-experiments do not. Sometimes, we also refer to non-experimental research, which is not really a research design, but an all-inclusive term that includes all types of research that do not employ treatment manipulation or random assignment, such as survey research, observational research, and correlational studies.

Basic Concepts

Treatment and control groups. In experimental research, some subjects are administered one or more experimental stimulus called a **treatment** (the **treatment group**) while other subjects are not given such a stimulus (the **control group**). The treatment may be considered successful if subjects in the treatment group rate more favorably on outcome variables than control group subjects. Multiple levels of experimental stimulus may be administered, in which case, there may be more than one treatment group. For example, in order to test the effects of a new drug intended to treat a certain medical condition like dementia, if a sample of dementia patients is randomly divided into three groups, with the first group receiving a high dosage of the drug, the second group receiving a low dosage, and the third group receives a placebo such as a sugar pill (control group), then the first two groups are experimental groups and the third group is a control group. After administering the drug for a period of time, if the condition of the experimental group subjects improved significantly more than the control group subjects, we can say that the drug is effective. We can also compare the conditions of the high and low dosage experimental groups to determine if the high dose is more effective than the low dose.

Treatment manipulation. Treatments are the unique feature of experimental research that sets this design apart from all other research methods. Treatment manipulation helps control for the "cause" in cause-effect relationships. Naturally, the validity of experimental research depends on how well the treatment was manipulated. Treatment manipulation must be checked using pretests and pilot tests prior to the experimental study. Any measurements conducted before the treatment is administered are called **pretest measures**, while those conducted after the treatment are **posttest measures**.

Random selection and assignment. Random selection is the process of randomly drawing a sample from a population or a sampling frame. This approach is typically employed in survey research, and assures that each unit in the population has a positive chance of being selected into the sample. Random assignment is however a process of randomly assigning subjects to experimental or control groups. This is a standard practice in true experimental research to ensure that treatment groups are similar (equivalent) to each other and to the control group, prior to treatment administration. Random selection is related to sampling, and is therefore, more closely related to the external validity (generalizability) of findings. However, random assignment is related to design, and is therefore most related to internal validity. It is possible to have both random selection and random assignment in well-designed experimental research, but quasi-experimental research involves neither random selection nor random assignment.

Threats to internal validity. Although experimental designs are considered more rigorous than other research methods in terms of the internal validity of their inferences (by virtue of their ability to control causes through treatment manipulation), they are not immune to internal validity threats. Some of these threats to internal validity are described below, within the context of a study of the impact of a special remedial math tutoring program for improving the math abilities of high school students.

- *History threat* is the possibility that the observed effects (dependent variables) are caused by extraneous or historical events rather than by the experimental treatment. For instance, students' post-remedial math score improvement may have been caused by their preparation for a math exam at their school, rather than the remedial math program.
- *Maturation threat* refers to the possibility that observed effects are caused by natural maturation of subjects (e.g., a general improvement in their intellectual ability to understand complex concepts) rather than the experimental treatment.
- *Testing threat* is a threat in pre-post designs where subjects' posttest responses are conditioned by their pretest responses. For instance, if students remember their answers from the pretest evaluation, they may tend to repeat them in the posttest exam. Not conducting a pretest can help avoid this threat.

- *Instrumentation threat*, which also occurs in pre-post designs, refers to the possibility that the difference between pretest and posttest scores is not due to the remedial math program, but due to changes in the administered test, such as the posttest having a higher or lower degree of difficulty than the pretest.
- *Mortality threat* refers to the possibility that subjects may be dropping out of the study at differential rates between the treatment and control groups due to a systematic reason, such that the dropouts were mostly students who scored low on the pretest. If the low-performing students drop out, the results of the posttest will be artificially inflated by the preponderance of high-performing students.
- *Regression threat*, also called a regression to the mean, refers to the statistical tendency of a group's overall performance on a measure during a posttest to regress toward the mean of that measure rather than in the anticipated direction. For instance, if subjects scored high on a pretest, they will have a tendency to score lower on the posttest (closer to the mean) because their high scores (away from the mean) during the pretest was possibly a statistical aberration. This problem tends to be more prevalent in non-random samples and when the two measures are imperfectly correlated.

Two-Group Experimental Designs

The simplest true experimental designs are two group designs involving one treatment group and one control group, and are ideally suited for testing the effects of a single independent variable that can be manipulated as a treatment. The two basic two-group designs are the pretest-posttest control group design and the posttest-only control group design, while variations may include covariance designs. These designs are often depicted using a standardized design notation, where R represents random assignment of subjects to groups, X represents the treatment administered to the treatment group, and O represents pretest or posttest observations of the dependent variable (with different subscripts to distinguish between pretest and posttest observations of treatment and control groups).

Pretest-posttest control group design. In this design, subjects are randomly assigned to treatment and control groups, subjected to an initial (pretest) measurement of the dependent variables of interest, the treatment group is administered a treatment (representing the independent variable of interest), and the dependent variables measured again (posttest). The notation of this design is shown in Figure 10.1.

$R \quad O_1 \quad X \quad O_2$ (Treatment group)
$R \quad O_3 \quad\quad\; O_4$ (Control group)

Figure 10.1. Pretest-posttest control group design

The effect *E* of the experimental treatment in the pretest posttest design is measured as the difference in the posttest and pretest scores between the treatment and control groups:

$$E = (O_2 - O_1) - (O_4 - O_3)$$

Statistical analysis of this design involves a simple analysis of variance (ANOVA) between the treatment and control groups. The pretest posttest design handles several threats to internal validity, such as maturation, testing, and regression, since these threats can be expected to influence both treatment and control groups in a similar (random) manner. The selection threat is controlled via random assignment. However, additional threats to internal validity may exist. For instance, mortality can be a problem if there are differential dropout rates between the two groups, and the pretest measurement may bias the posttest measurement (especially if the pretest introduces unusual topics or content).

Posttest-only control group design. This design is a simpler version of the pretest-posttest design where pretest measurements are omitted. The design notation is shown in Figure 10.2.

R	*X*	O_1	(Treatment group)
R		O_2	(Control group)

Figure 10.2. Posttest only control group design

The treatment effect is measured simply as the difference in the posttest scores between the two groups:

$$E = (O_1 - O_2)$$

The appropriate statistical analysis of this design is also a two-group analysis of variance (ANOVA). The simplicity of this design makes it more attractive than the pretest-posttest design in terms of internal validity. This design controls for maturation, testing, regression, selection, and pretest-posttest interaction, though the mortality threat may continue to exist.

Covariance designs. Sometimes, measures of dependent variables may be influenced by extraneous variables called *covariates*. Covariates are those variables that are not of central interest to an experimental study, but should nevertheless be controlled in an experimental design in order to eliminate their potential effect on the dependent variable and therefore allow for a more accurate detection of the effects of the independent variables of interest. The experimental designs discussed earlier did not control for such covariates. A covariance design (also called a concomitant variable design) is a special type of pretest posttest control group design where the pretest measure is essentially a measurement of the covariates of interest rather than that of the dependent variables. The design notation is shown in Figure 10.3, where C represents the covariates:

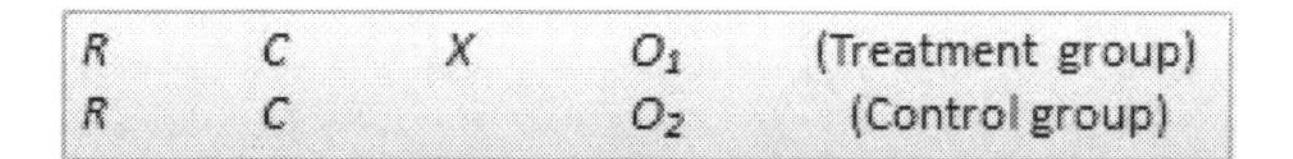

Figure 10.3. Covariance design

Because the pretest measure is not a measurement of the dependent variable, but rather a covariate, the treatment effect is measured as the difference in the posttest scores between the treatment and control groups as:

$$E = (O_1 - O_2)$$

Due to the presence of covariates, the right statistical analysis of this design is a two-group analysis of covariance (ANCOVA). This design has all the advantages of post-test only design, but with internal validity due to the controlling of covariates. Covariance designs can also be extended to pretest-posttest control group design.

Factorial Designs

Two-group designs are inadequate if your research requires manipulation of two or more independent variables (treatments). In such cases, you would need four or higher-group designs. Such designs, quite popular in experimental research, are commonly called factorial designs. Each independent variable in this design is called a *factor*, and each sub-division of a factor is called a *level*. Factorial designs enable the researcher to examine not only the individual effect of each treatment on the dependent variables (called main effects), but also their joint effect (called interaction effects).

The most basic factorial design is a 2 x 2 factorial design, which consists of two treatments, each with two levels (such as high/low or present/absent). For instance, let's say that you want to compare the learning outcomes of two different types of instructional techniques (in-class and online instruction), and you also want to examine whether these effects vary with the time of instruction (1.5 or 3 hours per week). In this case, you have two factors: instructional type and instructional time; each with two levels (in-class and online for instructional type, and 1.5 and 3 hours/week for instructional time), as shown in Figure 8.1. If you wish to add a third level of instructional time (say 6 hours/week), then the second factor will consist of three levels and you will have a 2 x 3 factorial design. On the other hand, if you wish to add a third factor such as group work (present versus absent), you will have a 2 x 2 x 2 factorial design. In this notation, each number represents a factor, and the value of each factor represents the number of levels in that factor.

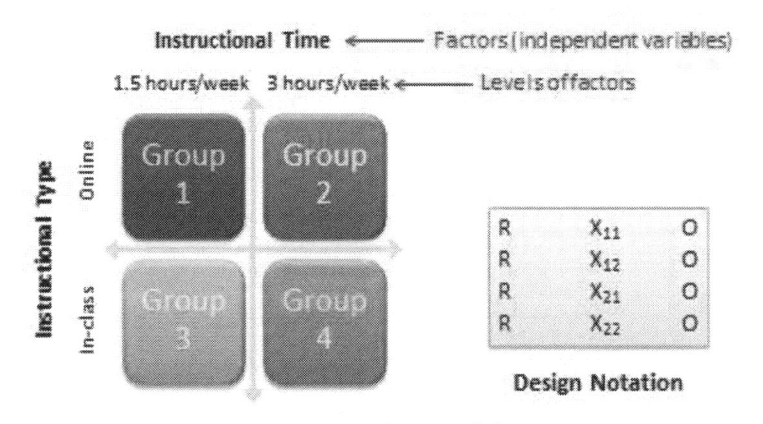

Figure 10.4. 2 x 2 factorial design

Factorial designs can also be depicted using a design notation, such as that shown on the right panel of Figure 10.4. R represents random assignment of subjects to treatment groups, X represents the treatment groups themselves (the subscripts of X represents the level of each factor), and O represent observations of the dependent variable. Notice that the 2 x 2 factorial design will have four treatment groups, corresponding to the four combinations of the two levels of each factor. Correspondingly, the 2 x 3 design will have six treatment groups, and the 2 x 2 x 2 design will have eight treatment groups. As a rule of thumb, each cell in a factorial design should have a minimum sample size of 20 (this estimate is derived from Cohen's power calculations based on medium effect sizes). So a 2 x 2 x 2 factorial design requires a minimum total sample size of 160 subjects, with at least 20 subjects in each cell. As you can see, the cost of data collection can increase substantially with more levels or factors in your factorial design. Sometimes, due to resource constraints, some cells in such factorial designs may not receive any treatment at all, which are called *incomplete factorial designs*. Such incomplete designs hurt our ability to draw inferences about the incomplete factors.

In a factorial design, a **main effect** is said to exist if the dependent variable shows a significant difference between multiple levels of one factor, at *all levels* of other factors. No change in the dependent variable across factor levels is the null case (baseline), from which main effects are evaluated. In the above example, you may see a main effect of instructional type, instructional time, or both on learning outcomes. An **interaction effect** exists when the effect of differences in one factor depends upon the level of a second factor. In our example, if the effect of instructional type on learning outcomes is greater for 3 hours/week of instructional time than for 1.5 hours/week, then we can say that there is an interaction effect between instructional type and instructional time on learning outcomes. Note that the presence of interaction effects dominate and make main effects irrelevant, and it is not meaningful to interpret main effects if interaction effects are significant.

Hybrid Experimental Designs

Hybrid designs are those that are formed by combining features of more established designs. Three such hybrid designs are randomized bocks design, Solomon four-group design, and switched replications design.

Randomized block design. This is a variation of the posttest-only or pretest-posttest control group design where the subject population can be grouped into relatively homogeneous subgroups (called *blocks*) within which the experiment is replicated. For instance, if you want to replicate the same posttest-only design among university students and full-time working professionals (two homogeneous blocks), subjects in both blocks are randomly split between treatment group (receiving the same treatment) or control group (see Figure 10.5). The purpose of this design is to reduce the "noise" or variance in data that may be attributable to differences between the blocks so that the actual effect of interest can be detected more accurately.

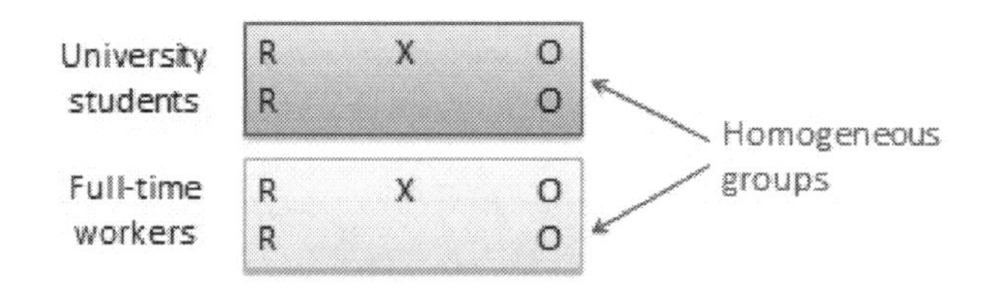

Figure 10.5. Randomized blocks design

Solomon four-group design. In this design, the sample is divided into two treatment groups and two control groups. One treatment group and one control group receive the pretest, and the other two groups do not. This design represents a combination of posttest-only and pretest-posttest control group design, and is intended to test for the potential biasing effect of pretest measurement on posttest measures that tends to occur in pretest-posttest designs but not in posttest only designs. The design notation is shown in Figure 10.6.

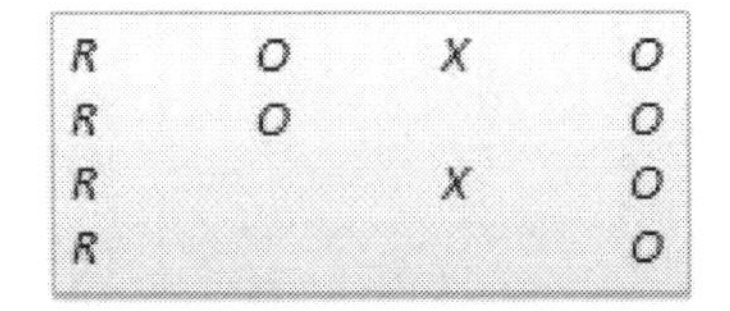

Figure 10.6. Solomon four-group design

Switched replication design. This is a two-group design implemented in two phases with three waves of measurement. The treatment group in the first phase serves as the control group in the second phase, and the control group in the first phase becomes the treatment group in the second phase, as illustrated in Figure 10.7. In other words, the original design is repeated or replicated temporally with treatment/control roles switched between the two groups. By the end of the study, all participants will have received the treatment either during the first or the second phase. This design is most feasible in organizational contexts where organizational programs (e.g., employee training) are implemented in a phased manner or are repeated at regular intervals.

Figure 10.7. Switched replication design

Quasi-Experimental Designs

Quasi-experimental designs are almost identical to true experimental designs, but lacking one key ingredient: random assignment. For instance, one entire class section or one organization is used as the treatment group, while another section of the same class or a different organization in the same industry is used as the control group. This lack of random assignment potentially results in groups that are non-equivalent, such as one group possessing greater mastery of a certain content than the other group, say by virtue of having a better teacher in a previous semester, which introduces the possibility of *selection bias*. Quasi-experimental designs are therefore inferior to true experimental designs in interval validity due to the presence of a variety of selection related threats such as selection-maturation threat (the treatment and control groups maturing at different rates), selection-history threat (the treatment and control groups being differentially impact by extraneous or historical events), selection-regression threat (the treatment and control groups regressing toward the mean between pretest and posttest at different rates), selection-instrumentation threat (the treatment and control groups responding differently to the measurement), selection-testing (the treatment and control groups responding differently to the pretest), and selection-mortality (the treatment and control groups demonstrating differential dropout rates). Given these selection threats, it is generally preferable to avoid quasi-experimental designs to the greatest extent possible.

Many true experimental designs can be converted to quasi-experimental designs by omitting random assignment. For instance, the quasi-equivalent version of pretest-posttest control group design is called **nonequivalent groups design** (NEGD), as shown in Figure 10.8, with random assignment *R* replaced by non-equivalent (non-random) assignment *N*. Likewise, the quasi-experimental version of switched replication design is called **non-equivalent switched replication design** (see Figure 10.9).

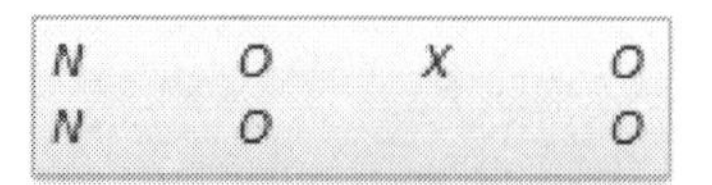

Figure 10.8. NEGD design

Figure 10.9. Non-equivalent switched replication design

In addition, there are quite a few unique non-equivalent designs without corresponding true experimental design cousins. Some of the more useful of these designs are discussed next.

Regression-discontinuity (RD) design. This is a non-equivalent pretest-posttest design where subjects are assigned to treatment or control group based on a cutoff score on a preprogram measure. For instance, patients who are severely ill may be assigned to a treatment group to test the efficacy of a new drug or treatment protocol and those who are mildly ill are assigned to the control group. In another example, students who are lagging behind on standardized test scores may be selected for a remedial curriculum program intended to improve their performance, while those who score high on such tests are not selected from the remedial program. The design notation can be represented as follows, where *C* represents the cutoff score:

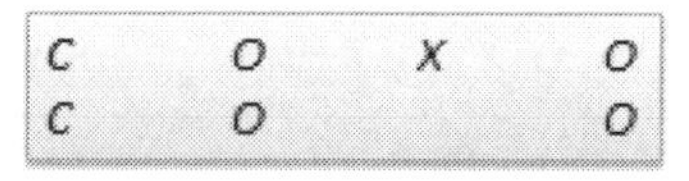

Figure 10.10. RD design

Because of the use of a cutoff score, it is possible that the observed results may be a function of the cutoff score rather than the treatment, which introduces a new threat to internal validity. However, using the cutoff score also ensures that limited or costly resources are distributed to people who need them the most rather than randomly across a population, while simultaneously allowing a quasi-experimental treatment. The control group scores in the RD design does not serve as a benchmark for comparing treatment group scores, given the systematic non-equivalence between the two groups. Rather, if there is no discontinuity between pretest and posttest scores in the control group, but such a discontinuity persists in the treatment group, then this discontinuity is viewed as evidence of the treatment effect.

Proxy pretest design. This design, shown in Figure 10.11, looks very similar to the standard NEGD (pretest-posttest) design, with one critical difference: the pretest score is collected after the treatment is administered. A typical application of this design is when a researcher is brought in to test the efficacy of a program (e.g., an educational program) after the program has already started and pretest data is not available. Under such circumstances, the best option for the researcher is often to use a different prerecorded measure, such as students' grade point average before the start of the program, as a proxy for pretest data. A variation of the proxy pretest design is to use subjects' posttest recollection of pretest data, which may be subject to recall bias, but nevertheless may provide a measure of *perceived* gain or change in the dependent variable.

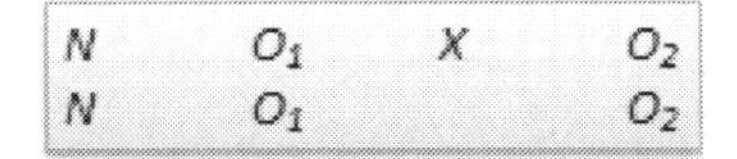

Figure 10.11. Proxy pretest design

Separate pretest-posttest samples design. This design is useful if it is not possible to collect pretest and posttest data from the same subjects for some reason. As shown in Figure 10.12, there are four groups in this design, but two groups come from a single non-equivalent group, while the other two groups come from a different non-equivalent group. For instance, you want to test customer satisfaction with a new online service that is implemented in one city but not in another. In this case, customers in the first city serve as the treatment group and those in the second city constitute the control group. If it is not possible to obtain pretest and posttest measures from the same customers, you can measure customer satisfaction at one point in time, implement the new service program, and measure customer satisfaction (with a different set of customers) after the program is implemented. Customer satisfaction is also measured in the control group at the same times as in the treatment group, but without the new program implementation. The design is not particularly strong, because you cannot examine the changes in any specific customer's satisfaction score before and after the implementation, but you can only examine *average* customer satisfaction scores. Despite the lower internal validity, this design may still be a useful way of collecting quasi-experimental data when pretest and posttest data are not available from the same subjects.

N_1	O		
N_1		X	O
N_2	O		
N_2			O

Figure 10.12. Separate pretest-posttest samples design

Nonequivalent dependent variable (NEDV) design. This is a single-group pre-post quasi-experimental design with two outcome measures, where one measure is theoretically expected to be influenced by the treatment and the other measure is not. For instance, if you are designing a new calculus curriculum for high school students, this curriculum is likely to influence students' posttest calculus scores but not algebra scores. However, the posttest algebra scores may still vary due to extraneous factors such as history or maturation. Hence, the pre-post algebra scores can be used as a control measure, while that of pre-post calculus can be treated as the treatment measure. The design notation, shown in Figure 10.13, indicates the single group by a single N, followed by pretest O_1 and posttest O_2 for calculus and algebra for the same group of students. This design is weak in internal validity, but its advantage lies in not having to use a separate control group.

An interesting variation of the NEDV design is a *pattern matching NEDV design*, which employs multiple outcome variables and a theory that explains how much each variable will be affected by the treatment. The researcher can then examine if the theoretical prediction is matched in actual observations. This *pattern-matching* technique, based on the degree of correspondence between theoretical and observed patterns is a powerful way of alleviating internal validity concerns in the original NEDV design.

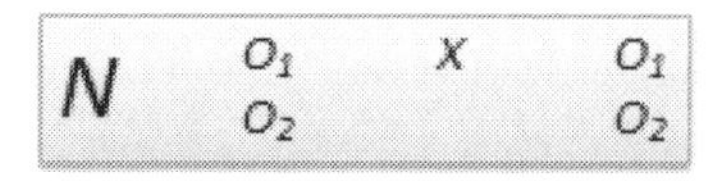

Figure 10.13. NEDV design

Perils of Experimental Research

Experimental research is one of the most difficult of research designs, and should not be taken lightly. This type of research is often best with a multitude of methodological problems. First, though experimental research requires theories for framing hypotheses for testing, much of current experimental research is atheoretical. Without theories, the hypotheses being tested tend to be ad hoc, possibly illogical, and meaningless. Second, many of the measurement instruments used in experimental research are not tested for reliability and validity, and are incomparable across studies. Consequently, results generated using such instruments are also incomparable. Third, many experimental research use inappropriate research designs, such as irrelevant dependent variables, no interaction effects, no experimental controls, and non-equivalent stimulus across treatment groups. Findings from such studies tend to lack internal validity and are highly suspect. Fourth, the treatments (tasks) used in experimental research may be diverse, incomparable, and inconsistent across studies and sometimes inappropriate for the subject population. For instance, undergraduate student subjects are often asked to pretend that they are marketing managers and asked to perform a complex budget allocation task in which they have no experience or expertise. The use of such inappropriate tasks, introduces new threats to internal validity (i.e., subject's performance may be an artifact of the content or difficulty of the task setting), generates findings that are non-interpretable and meaningless, and makes integration of findings across studies impossible.

The design of proper experimental treatments is a very important task in experimental design, because the treatment is the *raison d'etre* of the experimental method, and must never be rushed or neglected. To design an adequate and appropriate task, researchers should use prevalidated tasks if available, conduct treatment manipulation checks to check for the adequacy of such tasks (by debriefing subjects after performing the assigned task), conduct pilot tests (repeatedly, if necessary), and if doubt, using tasks that are simpler and familiar for the respondent sample than tasks that are complex or unfamiliar.

In summary, this chapter introduced key concepts in the experimental design research method and introduced a variety of true experimental and quasi-experimental designs. Although these designs vary widely in internal validity, designs with less internal validity should not be overlooked and may sometimes be useful under specific circumstances and empirical contingencies.

Chapter 11

Case Research

Case research, also called case study, is a method of intensively studying a phenomenon over time within its natural setting in one or a few sites. Multiple methods of data collection, such as interviews, observations, prerecorded documents, and secondary data, may be employed and inferences about the phenomenon of interest tend to be rich, detailed, and contextualized. Case research can be employed in a positivist manner for the purpose of theory testing or in an interpretive manner for theory building. This method is more popular in business research than in other social science disciplines.

Case research has several unique strengths over competing research methods such as experiments and survey research. First, case research can be used for either theory building or theory testing, while positivist methods can be used for theory testing only. In interpretive case research, the constructs of interest need not be known in advance, but may emerge from the data as the research progresses. Second, the research questions can be modified during the research process if the original questions are found to be less relevant or salient. This is not possible in any positivist method after the data is collected. Third, case research can help derive richer, more contextualized, and more authentic interpretation of the phenomenon of interest than most other research methods by virtue of its ability to capture a rich array of contextual data. Fourth, the phenomenon of interest can be studied from the perspectives of multiple participants and using multiple levels of analysis (e.g., individual and organizational).

At the same time, case research also has some inherent weaknesses. Because it involves no experimental control, internal validity of inferences remain weak. Of course, this is a common problem for all research methods except experiments. However, as described later, the problem of controls may be addressed in case research using "natural controls". Second, the quality of inferences derived from case research depends heavily on the integrative powers of the researcher. An experienced researcher may see concepts and patterns in case data that a novice researcher may miss. Hence, the findings are sometimes criticized as being subjective. Finally, because the inferences are heavily contextualized, it may be difficult to generalize inferences from case research to other contexts or other organizations.

It is important to recognize that case research is different from *case descriptions* such as Harvard case studies discussed in business classes. While case descriptions typically describe an organizational problem in rich detail with the goal of stimulating classroom discussion and critical thinking among students, or analyzing how well an organization handled a specific problem, case research is a formal research technique that involves a scientific method to derive explanations of organizational phenomena.

Case research is a difficult research method that requires advanced research skills on the part of the researcher, and is therefore, often prone to error. Benbasat et al. (1987)[8] describe five problems frequently encountered in case research studies. First, many case research studies start without specific research questions, and therefore end up without having any specific answers or insightful inferences. Second, case sites are often chosen based on access and convenience, rather than based on the fit with the research questions, and are therefore cannot adequately address the research questions of interest. Third, researchers often do not validate or triangulate data collected using multiple means, which may lead to biased interpretation based on responses from biased interviewees. Fourth, many studies provide very little details on how data was collected (e.g., what interview questions were used, which documents were examined, what are the organizational positions of each interviewee, etc.) or analyzed, which may raise doubts about the reliability of the inferences. Finally, despite its strength as a longitudinal research method, many case research studies do not follow through a phenomenon in a longitudinal manner, and hence present only a cross-sectional and limited view of organizational processes and phenomena that are temporal in nature.

Key Decisions in Case Research

Several key decisions must be made by a researcher when considering a case research method. First, is this the right method for the research questions being studied? The case research method is particularly appropriate for exploratory studies for discovering relevant constructs in areas where theory building at the formative stages, for studies where the experiences of participants and context of actions are critical, and for studies aimed at understanding complex, temporal processes (why and how of a phenomenon) rather than factors or causes (what). This method is well-suited for studying complex organizational processes that involve multiple participants and interacting sequences of events, such as organizational change and large-scale technology implementation projects.

Second, what is the appropriate unit of analysis for a case research study? Since case research can simultaneously examine multiple units of analyses, the researcher must decide whether she wishes to study a phenomenon at the individual, group, and organizational level or at multiple levels. For instance, a study of group decision making or group work may combine individual-level constructs such as individual participation in group activities with group-level constructs, such as group cohesion and group leadership, to derive richer understanding than that can be achieved from a single level of analysis.

Third, should the researcher employ a single-case or multiple-case design? The single case design is more appropriate at the outset of theory generation, if the situation is unique or extreme, if it is revelatory (i.e., the situation was previously inaccessible for scientific investigation), or if it represents a critical or contrary case for testing a well-formulated theory. The multiple case design is more appropriate for theory testing, for establishing generalizability of inferences, and for developing richer and more nuanced interpretations of a phenomenon. Yin (1984)[9] recommends the use of multiple case sites with replication logic, viewing each case site as similar to one experimental study, and following rules of scientific rigor similar to that used in positivist research.

---

[8] Benbasat, I., Goldstein, D. K., and Mead, M. (1987). "The Case Research Strategy in Studies of Information Systems," *MIS Quarterly* (11:3), 369-386.

[9] Yin, R. K. (2002), *Case Study Research: Design and Methods*. Thousand Oaks, CA: Sage Publications.

Fourth, what sites should be chosen for case research? Given the contextualized nature of inferences derived from case research, site selection is a particularly critical issue because selecting the wrong site may lead to the wrong inferences. If the goal of the research is to test theories or examine generalizability of inferences, then dissimilar case sites should be selected to increase variance in observations. For instance, if the goal of the research is to understand the process of technology implementation in firms, a mix of large, mid-sized, and small firms should be selected to examine whether the technology implementation process differs with firm size. Site selection should not be opportunistic or based on convenience, but rather based on the fit with research questions though a process called "theoretical sampling."

Fifth, what techniques of data collection should be used in case research? Although interview (either open-ended/unstructured or focused/structured) is by far the most popular data collection technique for case research, interview data can be supplemented or corroborated with other techniques such as direct observation (e.g., attending executive meetings, briefings, and planning sessions), documentation (e.g., internal reports, presentations, and memoranda, as well as external accounts such as newspaper reports), archival records (e.g., organization charts, financial records, etc.), and physical artifacts (e.g., devices, outputs, tools). Furthermore, the researcher should triangulate or validate observed data by comparing responses between interviewees.

Conducting Case Research

Most case research studies tend to be interpretive in nature. Interpretive case research is an inductive technique where evidence collected from one or more case sites is systematically analyzed and synthesized to allow concepts and patterns to emerge for the purpose of building new theories or expanding existing ones. Eisenhardt (1989)[10] propose a "roadmap" for building theories from case research, a slightly modified version of which is described below. For positivist case research, some of the following stages may need to be rearranged or modified; however sampling, data collection, and data analytic techniques should generally remain the same.

Define research questions. Like any other scientific research, case research must also start with defining research questions that are theoretically and practically interesting, and identifying some intuitive expectations about possible answers to those research questions or preliminary constructs to guide initial case design. In positivist case research, the preliminary constructs are based on theory, while no such theory or hypotheses should be considered *ex ante* in interpretive research. These research questions and constructs may be changed in interpretive case research later on, if needed, but not in positivist case research.

Select case sites. The researcher should use a process of "theoretical sampling" (not random sampling) to identify case sites. In this approach, case sites are chosen based on theoretical, rather than statistical, considerations, for instance, to replicate previous cases, to extend preliminary theories, or to fill theoretical categories or polar types. Care should be taken to ensure that the selected sites fit the nature of research questions, minimize extraneous variance or noise due to firm size, industry effects, and so forth, and maximize variance in the dependent variables of interest. For instance, if the goal of the research is to examine how some firms innovate better than others, the researcher should select firms of similar size within the

---

[10] Eisenhardt, K. M. (1989). "Building Theories from Case Research," *Academy of Management Review* (14:4), 532-550.

same industry to reduce industry or size effects, and select some more innovative and some less innovative firms to increase variation in firm innovation. Instead of cold-calling or writing to a potential site, it is better to contact someone at executive level inside each firm who has the authority to approve the project or someone who can identify a person of authority. During initial conversations, the researcher should describe the nature and purpose of the project, any potential benefits to the case site, how the collected data will be used, the people involved in data collection (other researchers, research assistants, etc.), desired interviewees, and the amount of time, effort, and expense required of the sponsoring organization. The researcher must also assure confidentiality, privacy, and anonymity of both the firm and the individual respondents.

Create instruments and protocols. Since the primary mode of data collection in case research is interviews, an interview protocol should be designed to guide the interview process. This is essentially a list of questions to be asked. Questions may be open-ended (unstructured) or closed-ended (structured) or a combination of both. The interview protocol must be strictly followed, and the interviewer must not change the order of questions or skip any question during the interview process, although some deviations are allowed to probe further into respondent's comments that are ambiguous or interesting. The interviewer must maintain a neutral tone, not lead respondents in any specific direction, say by agreeing or disagreeing with any response. More detailed interviewing techniques are discussed in the chapter on surveys. In addition, additional sources of data, such as internal documents and memorandums, annual reports, financial statements, newspaper articles, and direct observations should be sought to supplement and validate interview data.

Select respondents. Select interview respondents at different organizational levels, departments, and positions to obtain divergent perspectives on the phenomenon of interest. A random sampling of interviewees is most preferable; however a *snowball sample* is acceptable, as long as a diversity of perspectives is represented in the sample. Interviewees must be selected based on their personal involvement with the phenomenon under investigation and their ability and willingness to answer the researcher's questions accurately and adequately, and not based on convenience or access.

Start data collection. It is usually a good idea to electronically record interviews for future reference. However, such recording must only be done with the interviewee's consent. Even when interviews are being recorded, the interviewer should take notes to capture important comments or critical observations, behavioral responses (e.g., respondent's body language), and the researcher's personal impressions about the respondent and his/her comments. After each interview is completed, the entire interview should be transcribed verbatim into a text document for analysis.

Conduct within-case data analysis. Data analysis may follow or overlap with data collection. Overlapping data collection and analysis has the advantage of adjusting the data collection process based on themes emerging from data analysis, or to further probe into these themes. Data analysis is done in two stages. In the first stage (within-case analysis), the researcher should examine emergent concepts separately at each case site and patterns between these concepts to generate an initial theory of the problem of interest. The researcher can interview data subjectively to "make sense" of the research problem in conjunction with using her personal observations or experience at the case site. Alternatively, a coding strategy such as Glasser and Strauss' (1967) grounded theory approach, using techniques such as open coding, axial coding, and selective coding, may be used to derive a chain of evidence and inferences. These techniques are discussed in detail in a later chapter. Homegrown techniques, such as graphical representation of data (e.g., network diagram) or sequence analysis (for longitudinal data) may also be used. Note that there is no predefined way of analyzing the various types of case data, and the data analytic techniques can be modified to fit the nature of the research project.

Conduct cross-case analysis. Multi-site case research requires cross-case analysis as the second stage of data analysis. In such analysis, the researcher should look for similar concepts and patterns between different case sites, ignoring contextual differences that may lead to idiosyncratic conclusions. Such patterns may be used for validating the initial theory, or for refining it (by adding or dropping concepts and relationships) to develop a more inclusive and generalizable theory. This analysis may take several forms. For instance, the researcher may select categories (e.g., firm size, industry, etc.) and look for within-group similarities and between-group differences (e.g., high versus low performers, innovators versus laggards). Alternatively, she can compare firms in a pair-wise manner listing similarities and differences across pairs of firms.

Build and test hypotheses. Based on emergent concepts and themes that are generalizable across case sites, tentative hypotheses are constructed. These hypotheses should be compared iteratively with observed evidence to see if they fit the observed data, and if not, the constructs or relationships should be refined. Also the researcher should compare the emergent constructs and hypotheses with those reported in the prior literature to make a case for their internal validity and generalizability. Conflicting findings must not be rejected, but rather reconciled using creative thinking to generate greater insight into the emergent theory. When further iterations between theory and data yield no new insights or changes in the existing theory, "theoretical saturation" is reached and the theory building process is complete.

Write case research report. In writing the report, the researcher should describe very clearly the detailed process used for sampling, data collection, data analysis, and hypotheses development, so that readers can independently assess the reasonableness, strength, and consistency of the reported inferences. A high level of clarity in research methods is needed to ensure that the findings are not biased by the researcher's preconceptions.

Interpretive Case Research Exemplar

Perhaps the best way to learn about interpretive case research is to examine an illustrative example. One such example is Eisenhardt's (1989)[11] study of how executives make decisions in high-velocity environments (HVE). Readers are advised to read the original paper published in *Academy of Management Journal* before reading the synopsis in this chapter. In this study, Eisenhardt examined how executive teams in some HVE firms make fast decisions, while those in other firms cannot, and whether faster decisions improve or worsen firm performance in such environments. HVE was defined as one where demand, competition, and technology changes so rapidly and discontinuously that the information available is often inaccurate, unavailable or obsolete. The implicit assumptions were that (1) it is hard to make fast decisions with inadequate information in HVE, and (2) fast decisions may not be efficient and may result in poor firm performance.

---

[11] Eisenhardt, K. M. (1989). "Making Fast Strategic Decisions in High-Velocity Environments," *Academy of Management Journal* (32:3), 543-576.

Reviewing the prior literature on executive decision-making, Eisenhardt found several patterns, although none of these patterns were specific to high-velocity environments. The literature suggested that in the interest of expediency, firms that make faster decisions obtain input from fewer sources, consider fewer alternatives, make limited analysis, restrict user participation in decision-making, centralize decision-making authority, and has limited internal conflicts. However, Eisenhardt contended that these views may not necessarily explain how decision makers make decisions in high-velocity environments, where decisions must be made quickly and with incomplete information, while maintaining high decision quality.

To examine this phenomenon, Eisenhardt conducted an inductive study of eight firms in the personal computing industry. The personal computing industry was undergoing dramatic changes in technology with the introduction of the UNIX operating system, RISC architecture, and 64KB random access memory in the 1980's, increased competition with the entry of IBM into the personal computing business, and growing customer demand with double-digit demand growth, and therefore fit the profile of the high-velocity environment. This was a multiple case design with replication logic, where each case was expected to confirm or disconfirm inferences from other cases. Case sites were selected based on their access and proximity to the researcher; however, all of these firms operated in the high-velocity personal computing industry in California's Silicon Valley area. The collocation of firms in the same industry and the same area ruled out any "noise" or variance in dependent variables (decision speed or performance) attributable to industry or geographic differences.

The study employed an embedded design with multiple levels of analysis: decision (comparing multiple strategic decisions within each firm), executive teams (comparing different teams responsible for strategic decisions), and the firm (overall firm performance). Data was collected from five sources:

- Initial interviews with Chief Executive Officers: CEOs were asked questions about their firm's competitive strategy, distinctive competencies, major competitors, performance, and recent/ongoing major strategic decisions. Based on these interviews, several strategic decisions were selected in each firm for further investigation. Four criteria were used to select decisions: (1) the decisions involved the firm's strategic positioning, (2) the decisions had high stakes, (3) the decisions involved multiple functions, and (4) the decisions were representative of strategic decision-making process in that firm.
- Interviews with divisional heads: Each divisional head was asked sixteen open-ended questions, ranging from their firm's competitive strategy, functional strategy, top management team members, frequency and nature of interaction with team, typical decision making processes, how each of the previously identified decision was made, and how long it took them to make those decisions. Interviews lasted between 1.5 and 2 hours, and sometimes extended to 4 hours. To focus on facts and actual events rather than respondents' perceptions or interpretations, a "courtroom" style questioning was employed, such as when did this happen, what did you do, etc. Interviews were conducted by two people, and the data was validated by cross-checking facts and impressions made by the interviewer and note-taker. All interview data was recorded, however notes were also taken during each interview, which ended with the interviewer's overall impressions. Using a "24-hour rule", detailed field notes were completed within 24 hours of the interview, so that some data or impressions were not lost to recall.

- Questionnaires: Executive team members at each firm were completed a survey questionnaire that captured quantitative data on the extent of conflict and power distribution in their firm.
- Secondary data: Industry reports and internal documents such as demographics of the executive teams (responsible for strategic decisions), financial performance of firms, and so forth, were examined.
- Personal observation: Lastly, the researcher attended a 1-day strategy session and a weekly executive meeting at two firms in her sample.

Data analysis involved a combination of quantitative and qualitative techniques. Quantitative data on conflict and power were analyzed for patterns across firms/decisions. Qualitative interview data was combined into decision climate profiles, using profile traits (e.g., impatience) mentioned by more than one executive. For within-case analysis, decision stories were created for each strategic decision by combining executive accounts of the key decision events into a timeline. For cross-case analysis, pairs of firms were compared for similarities and differences, categorized along variables of interest such as decision speed and firm performance. Based on these analyses, tentative constructs and propositions were derived inductively from each decision story within firm categories. Each decision case was revisited to confirm the proposed relationships. The inferred propositions were compared with findings from the existing literature to reconcile examine differences with the extant literature and to generate new insights from the case findings. Finally, the validated propositions were synthesized into an inductive theory of strategic decision-making by firms in high-velocity environments.

Inferences derived from this multiple case research contradicted several decision-making patterns expected from the existing literature. First, fast decision makers in high-velocity environments used more information, and not less information as suggested by the previous literature. However, these decision makers used more *real-time* information (an insight not available from prior research), which helped them identify and respond to problems, opportunities, and changing circumstances faster. Second, fast decision makers examined more (not fewer) alternatives. However, they considered these multiple alternatives in a *simultaneous* manner, while slower decision makers examined fewer alternatives in a *sequential* manner. Third, fast decision makers did not centralize decision making or restrict inputs from others, as the literature suggested. Rather, these firms used a two-tiered decision process in which experienced counselors were asked for inputs in the first stage, following by a rapid comparison and decision selection in the second stage. Fourth, fast decision makers did not have less conflict, as expected from the literature, but employed better conflict resolution techniques to reduce conflict and improve decision-making speed. Finally, fast decision makers exhibited superior firm performance by virtue of their built-in cognitive, emotional, and political processes that led to rapid closure of major decisions.

Positivist Case Research Exemplar

Case research can also be used in a positivist manner to test theories or hypotheses. Such studies are rare, but Markus (1983)[12] provides an exemplary illustration in her study of technology implementation at the Golden Triangle Company (a pseudonym). The goal of this study was to understand why a newly implemented financial information system (FIS),

---

[12] Markus, M. L. (1983). "Power, Politics, and MIS Implementation," *Communications of the ACM* (26:6), 430-444.

intended to improve the productivity and performance of accountants at GTC was supported by accountants at GTC's corporate headquarters but resisted by divisional accountants at GTC branches. Given the uniqueness of the phenomenon of interest, this was a single-case research study.

To explore the reasons behind user resistance of FIS, Markus posited three alternative explanations: (1) system-determined theory: resistance was caused by factors related to an inadequate system, such as its technical deficiencies, poor ergonomic design, or lack of user friendliness, (2) people-determined theory: resistance was caused by factors internal to users, such as the accountants' cognitive styles or personality traits that were incompatible with using the system, and (3) interaction theory: resistance was not caused not by factors intrinsic to the system or the people, but by the interaction between the two set of factors. Specifically, interaction theory suggested that the FIS engendered a redistribution of intra-organizational power, and accountants who lost organizational status, relevance, or power as a result of FIS implementation resisted the system while those gaining power favored it.

In order to test the three theories, Markus predicted alternative outcomes expected from each theoretical explanation and analyzed the extent to which those predictions matched with her observations at GTC. For instance, the system-determined theory suggested that since user resistance was caused by an inadequate system, fixing the technical problems of the system would eliminate resistance. The computer running the FIS system was subsequently upgraded with a more powerful operating system, online processing (from initial batch processing, which delayed immediate processing of accounting information), and a simplified software for new account creation by managers. One year after these changes were made, the resistant users were still resisting the system and felt that it should be replaced. Hence, the system-determined theory was rejected.

The people-determined theory predicted that replacing individual resistors or co-opting them with less resistant users would reduce their resistance toward the FIS. Subsequently, GTC started a job rotation and mobility policy, moving accountants in and out of the resistant divisions, but resistance not only persisted, but in some cases increased! In one specific instance, one accountant, who was one of the system's designers and advocates when he worked for corporate accounting, started resisting the system after he was moved to the divisional controller's office. Failure to realize the predictions of the people-determined theory led to the rejection of this theory.

Finally, the interaction theory predicted that neither changing the system or the people (i.e., user education or job rotation policies) will reduce resistance as long as the power imbalance and redistribution from the pre-implementation phase were not addressed. Before FIS implementation, divisional accountants at GTC felt that they owned all accounting data related to their divisional operations. They maintained this data in thick, manual ledger books, controlled others' access to the data, and could reconcile unusual accounting events before releasing those reports. Corporate accountants relied heavily on divisional accountants for access to the divisional data for corporate reporting and consolidation. Because the FIS system automatically collected all data at source and consolidated them into a single corporate database, it obviated the need for divisional accountants, loosened their control and autonomy over their division's accounting data, and making their job somewhat irrelevant. Corporate accountants could now query the database and access divisional data directly without going through the divisional accountants, analyze and compare the performance of individual divisions, and report unusual patterns and activities to the executive committee, resulting in further erosion of the divisions' power. Though Markus did not empirically test this theory, her observations about the redistribution of organizational power, coupled with the rejection of the two alternative theories, led to the justification of interaction theory.

Comparisons with Traditional Research

Positivist case research, aimed at hypotheses testing, is often criticized by natural science researchers as lacking in controlled observations, controlled deductions, replicability, and generalizability of findings - the traditional principles of positivist research. However, these criticisms can be overcome through appropriate case research designs. For instance, the problem of controlled observations refers to the difficulty of obtaining experimental or statistical control in case research. However, case researchers can compensate for such lack of controls by employing "natural controls." This natural control in Markus' (1983) study was the corporate accountant who was one of the system advocates initially, but started resisting it once he moved to controlling division. In this instance, the change in his behavior may be attributed to his new divisional position. However, such natural controls cannot be anticipated in advance, and case researchers may overlook then unless they are proactively looking for such controls. Incidentally, natural controls are also used in natural science disciplines such as astronomy, geology, and human biology, such as wait for comets to pass close enough to the earth in order to make inferences about comets and their composition.

The problem of controlled deduction refers to the lack of adequate quantitative evidence to support inferences, given the mostly qualitative nature of case research data. Despite the lack of quantitative data for hypotheses testing (e.g., t-tests), controlled deductions can still be obtained in case research by generating behavioral predictions based on theoretical considerations and testing those predictions over time. Markus employed this strategy in her study by generating three alternative theoretical hypotheses for user resistance, and rejecting two of those predictions when they did not match with actual observed behavior. In this case, the hypotheses were tested using logical propositions rather than using mathematical tests, which are just as valid as statistical inferences since mathematics is a subset of logic.

Third, the problem of replicability refers to the difficulty of observing the same phenomenon given the uniqueness and idiosyncrasy of a given case site. However, using Markus' three theories as an illustration, a different researcher can test the same theories at a different case site, where three different predictions may emerge based on the idiosyncratic nature of the new case site, and the three resulting predictions may be tested accordingly. In other words, it is possible to replicate the inferences of case research, even if the case research site or context may not be replicable.

Fourth, case research tends to examine unique and non-replicable phenomena that may not be generalized to other settings. Generalizability in natural sciences is established through additional studies. Likewise, additional case studies conducted in different contexts with different predictions can establish generalizability of findings if such findings are observed to be consistent across studies.

Lastly, British philosopher Karl Popper described four requirements of scientific theories: (1) theories should be falsifiable, (2) they should be logically consistent, (3) they should have adequate predictive ability, and (4) they should provide better explanation than rival theories. In case research, the first three requirements can be increased by increasing the degrees of freedom of observed findings, such as by increasing the number of case sites, the number of alternative predictions, and the number of levels of analysis examined. This was accomplished in Markus' study by examining the behavior of multiple groups (divisional accountants and corporate accountants) and providing multiple (three) rival explanations. Popper's fourth condition was accomplished in this study when one hypothesis was found to match observed evidence better than the two rival hypotheses.

Chapter 12

Interpretive Research

The last chapter introduced interpretive research, or more specifically, interpretive case research. This chapter will explore other kinds of interpretive research. Recall that positivist or deductive methods, such as laboratory experiments and survey research, are those that are specifically intended for theory (or hypotheses) testing, while interpretive or inductive methods, such as action research and ethnography, are intended for theory building. Unlike a positivist method, where the researcher starts with a theory and tests theoretical postulates using empirical data, in interpretive methods, the researcher starts with data and tries to derive a theory about the phenomenon of interest from the observed data.

The term "interpretive research" is often used loosely and synonymously with "qualitative research", although the two concepts are quite different. Interpretive research is a research paradigm (see Chapter 3) that is based on the assumption that social reality is not singular or objective, but is rather shaped by human experiences and social contexts (ontology), and is therefore best studied within its socio-historic context by reconciling the subjective interpretations of its various participants (epistemology). Because interpretive researchers view social reality as being embedded within and impossible to abstract from their social settings, they "interpret" the reality though a "sense-making" process rather than a hypothesis testing process. This is in contrast to the positivist or functionalist paradigm that assumes that the reality is relatively independent of the context, can be abstracted from their contexts, and studied in a decomposable functional manner using objective techniques such as standardized measures. Whether a researcher should pursue interpretive or positivist research depends on paradigmatic considerations about the nature of the phenomenon under consideration and the best way to study it.

However, qualitative versus quantitative research refers to empirical or data-oriented considerations about the type of data to collect and how to analyze them. Qualitative research relies mostly on non-numeric data, such as interviews and observations, in contrast to quantitative research which employs numeric data such as scores and metrics. Hence, qualitative research is not amenable to statistical procedures such as regression analysis, but is coded using techniques like content analysis. Sometimes, coded qualitative data is tabulated quantitatively as frequencies of codes, but this data is not statistically analyzed. Many puritan interpretive researchers reject this coding approach as a futile effort to seek consensus or objectivity in a social phenomenon which is essentially subjective.

Although interpretive research tends to rely heavily on qualitative data, quantitative data may add more precision and clearer understanding of the phenomenon of interest than qualitative data. For example, Eisenhardt (1989), in her interpretive study of decision making n high-velocity firms (discussed in the previous chapter on case research), collected numeric data on how long it took each firm to make certain strategic decisions (which ranged from 1.5 months to 18 months), how many decision alternatives were considered for each decision, and surveyed her respondents to capture their perceptions of organizational conflict. Such numeric data helped her clearly distinguish the high-speed decision making firms from the low-speed decision makers, without relying on respondents' subjective perceptions, which then allowed her to examine the number of decision alternatives considered by and the extent of conflict in high-speed versus low-speed firms. Interpretive research should attempt to collect both qualitative and quantitative data pertaining to their phenomenon of interest, and so should positivist research as well. Joint use of qualitative and quantitative data, often called "mixed-mode designs", may lead to unique insights and are highly prized in the scientific community.

Interpretive research has its roots in anthropology, sociology, psychology, linguistics, and semiotics, and has been available since the early 19th century, long before positivist techniques were developed. Many positivist researchers view interpretive research as erroneous and biased, given the subjective nature of the qualitative data collection and interpretation process employed in such research. However, the failure of many positivist techniques to generate interesting insights or new knowledge have resulted in a resurgence of interest in interpretive research since the 1970's, albeit with exacting methods and stringent criteria to ensure the reliability and validity of interpretive inferences.

Distinctions from Positivist Research

In addition to fundamental paradigmatic differences in ontological and epistemological assumptions discussed above, interpretive and positivist research differ in several other ways. First, interpretive research employs a *theoretical sampling* strategy, where study sites, respondents, or cases are selected based on theoretical considerations such as whether they fit the phenomenon being studied (e.g., sustainable practices can only be studied in organizations that have implemented sustainable practices), whether they possess certain characteristics that make them uniquely suited for the study (e.g., a study of the drivers of firm innovations should include some firms that are high innovators and some that are low innovators, in order to draw contrast between these firms), and so forth. In contrast, positivist research employs *random sampling* (or a variation of this technique), where cases are chosen randomly from a population, for purposes of generalizability. Hence, convenience samples and small samples are considered acceptable in interpretive research as long as they fit the nature and purpose of the study, but not in positivist research.

Second, the role of the researcher receives critical attention in interpretive research. In some methods such as ethnography, action research, and participant observation, the researcher is considered part of the social phenomenon, and her specific role and involvement in the research process must be made clear during data analysis. In other methods, such as case research, the researcher must take a "neutral" or unbiased stance during the data collection and analysis processes, and ensure that her personal biases or preconceptions does not taint the nature of subjective inferences derived from interpretive research. In positivist research, however, the researcher is considered to be external to and independent of the research context and is not presumed to bias the data collection and analytic procedures.

Third, interpretive analysis is holistic and contextual, rather than being reductionist and isolationist. Interpretive interpretations tend to focus on language, signs, and meanings from the perspective of the participants involved in the social phenomenon, in contrast to statistical techniques that are employed heavily in positivist research. Rigor in interpretive research is viewed in terms of systematic and transparent approaches for data collection and analysis rather than statistical benchmarks for construct validity or significance testing.

Lastly, data collection and analysis can proceed simultaneously and iteratively in interpretive research. For instance, the researcher may conduct an interview and code it before proceeding to the next interview. Simultaneous analysis helps the researcher correct potential flaws in the interview protocol or adjust it to capture the phenomenon of interest better. The researcher may even change her original research question if she realizes that her original research questions are unlikely to generate new or useful insights. This is a valuable but often understated benefit of interpretive research, and is not available in positivist research, where the research project cannot be modified or changed once the data collection has started without redoing the entire project from the start.

Benefits and Challenges of Interpretive Research

Interpretive research has several unique advantages. First, they are well-suited for exploring hidden reasons behind complex, interrelated, or multifaceted social processes, such as inter-firm relationships or inter-office politics, where quantitative evidence may be biased, inaccurate, or otherwise difficult to obtain. Second, they are often helpful for theory construction in areas with no or insufficient a priori theory. Third, they are also appropriate for studying context-specific, unique, or idiosyncratic events or processes. Fourth, interpretive research can also help uncover interesting and relevant research questions and issues for follow-up research.

At the same time, interpretive research also has its own set of challenges. First, this type of research tends to be more time and resource intensive than positivist research in data collection and analytic efforts. Too little data can lead to false or premature assumptions, while too much data may not be effectively processed by the researcher. Second, interpretive research requires well-trained researchers who are capable of seeing and interpreting complex social phenomenon from the perspectives of the embedded participants and reconciling the diverse perspectives of these participants, without injecting their personal biases or preconceptions into their inferences. Third, all participants or data sources may not be equally credible, unbiased, or knowledgeable about the phenomenon of interest, or may have undisclosed political agendas, which may lead to misleading or false impressions. Inadequate trust between participants and researcher may hinder full and honest self-representation by participants, and such trust building takes time. It is the job of the interpretive researcher to "see through the smoke" (hidden or biased agendas) and understand the true nature of the problem. Fourth, given the heavily contextualized nature of inferences drawn from interpretive research, such inferences do not lend themselves well to replicability or generalizability. Finally, interpretive research may sometimes fail to answer the research questions of interest or predict future behaviors.

Characteristics of Interpretive Research

All interpretive research must adhere to a common set of principles, as described below.

Naturalistic inquiry: Social phenomena must be studied within their natural setting. Because interpretive research assumes that social phenomena are situated within and cannot be isolated from their social context, interpretations of such phenomena must be grounded within their socio-historical context. This implies that contextual variables should be observed and considered in seeking explanations of a phenomenon of interest, even though context sensitivity may limit the generalizability of inferences.

Researcher as instrument: Researchers are often embedded within the social context that they are studying, and are considered part of the data collection instrument in that they must use their observational skills, their trust with the participants, and their ability to extract the correct information. Further, their personal insights, knowledge, and experiences of the social context is critical to accurately interpreting the phenomenon of interest. At the same time, researchers must be fully aware of their personal biases and preconceptions, and not let such biases interfere with their ability to present a fair and accurate portrayal of the phenomenon.

Interpretive analysis: Observations must be interpreted through the eyes of the participants embedded in the social context. Interpretation must occur at two levels. The first level involves viewing or experiencing the phenomenon from the subjective perspectives of the social participants. The second level is to understand the meaning of the participants' experiences in order to provide a "thick description" or a rich narrative story of the phenomenon of interest that can communicate why participants acted the way they did.

Use of expressive language: Documenting the verbal and non-verbal language of participants and the analysis of such language are integral components of interpretive analysis. The study must ensure that the story is viewed through the eyes of a person, and not a machine, and must depict the emotions and experiences of that person, so that readers can understand and relate to that person. Use of imageries, metaphors, sarcasm, and other figures of speech is very common in interpretive analysis.

Temporal nature: Interpretive research is often not concerned with searching for specific answers, but with understanding or "making sense of" a dynamic social process as it unfolds over time. Hence, such research requires an immersive involvement of the researcher at the study site for an extended period of time in order to capture the entire evolution of the phenomenon of interest.

Hermeneutic circle: Interpretive interpretation is an iterative process of moving back and forth from pieces of observations (text) to the entirety of the social phenomenon (context) to reconcile their apparent discord and to construct a theory that is consistent with the diverse subjective viewpoints and experiences of the embedded participants. Such iterations between the understanding/meaning of a phenomenon and observations must continue until "theoretical saturation" is reached, whereby any additional iteration does not yield any more insight into the phenomenon of interest.

Interpretive Data Collection

Data is collected in interpretive research using a variety of techniques. The most frequently used technique is **interviews** (face-to-face, telephone, or focus groups). Interview types and strategies are discussed in detail in a previous chapter on survey research. A second technique is **observation**. Observational techniques include *direct observation*, where the researcher is a neutral and passive external observer and is not involved in the phenomenon of interest (as in case research), and *participant observation*, where the researcher is an active participant in the phenomenon and her inputs or mere presence influence the phenomenon being studied (as in action research). A third technique is **documentation**, where external and internal documents, such as memos, electronic mails, annual reports, financial statements, newspaper articles, websites, may be used to cast further insight into the phenomenon of interest or to corroborate other forms of evidence.

Interpretive Research Designs

Case research. As discussed in the previous chapter, case research is an intensive longitudinal study of a phenomenon at one or more research sites for the purpose of deriving detailed, contextualized inferences and understanding the dynamic process underlying a phenomenon of interest. Case research is a unique research design in that it can be used in an interpretive manner to build theories or in a positivist manner to test theories. The previous chapter on case research discusses both techniques in depth and provides illustrative exemplars. Furthermore, the case researcher is a neutral observer (direct observation) in the social setting rather than an active participant (participant observation). As with any other interpretive approach, drawing meaningful inferences from case research depends heavily on the observational skills and integrative abilities of the researcher.

Action research. Action research is a qualitative but positivist research design aimed at theory testing rather than theory building (discussed in this chapter due to lack of a proper space). This is an interactive design that assumes that complex social phenomena are best understood by introducing changes, interventions, or "actions" into those phenomena and observing the outcomes of such actions on the phenomena of interest. In this method, the researcher is usually a consultant or an organizational member embedded into a social context (such as an organization), who initiates an action in response to a social problem, and examines how her action influences the phenomenon while also learning and generating insights about the relationship between the action and the phenomenon. Examples of actions may include organizational change programs, such as the introduction of new organizational processes, procedures, people, or technology or replacement of old ones, initiated with the goal of improving an organization's performance or profitability in its business environment. The researcher's choice of actions must be based on theory, which should explain why and how such actions may bring forth the desired social change. The theory is validated by the extent to which the chosen action is successful in remedying the targeted problem. Simultaneous problem solving and insight generation is the central feature that distinguishes action research from other research methods (which may not involve problem solving) and from consulting (which may not involve insight generation). Hence, action research is an excellent method for bridging research and practice.

There are several variations of the action research method. The most popular of these method is the participatory action research, designed by Susman and Evered (1978)[13]. This method follows an action research cycle consisting of five phases: (1) diagnosing, (2) action planning, (3) action taking, (4) evaluating, and (5) learning (see Figure 10.1). *Diagnosing* involves identifying and defining a problem in its social context. *Action planning* involves identifying and evaluating alternative solutions to the problem, and deciding on a future course of action (based on theoretical rationale). *Action taking* is the implementation of the planned course of action. The *evaluation* stage examines the extent to which the initiated action is

---

[13] Susman, G.I. and Evered, R.D. (1978). "An Assessment of the Scientific Merits of Action Research," *Administrative Science Quarterly*, (23), 582-603.

successful in resolving the original problem, i.e., whether theorized effects are indeed realized in practice. In the *learning* phase, the experiences and feedback from action evaluation are used to generate insights about the problem and suggest future modifications or improvements to the action. Based on action evaluation and learning, the action may be modified or adjusted to address the problem better, and the action research cycle is repeated with the modified action sequence. It is suggested that the entire action research cycle be traversed at least twice so that learning from the first cycle can be implemented in the second cycle. The primary mode of data collection is participant observation, although other techniques such as interviews and documentary evidence may be used to corroborate the researcher's observations.

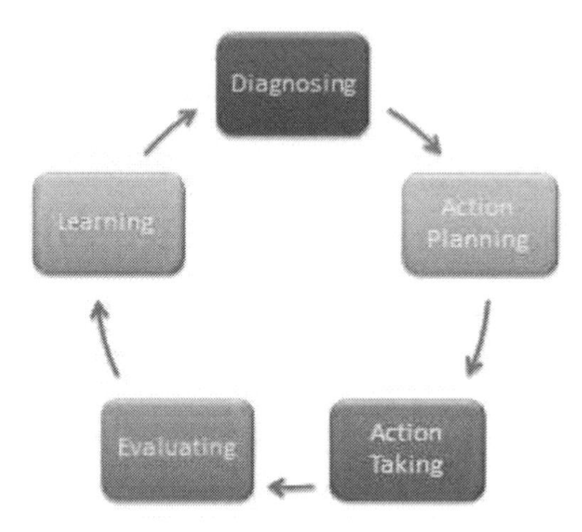

Figure 10.1. Action research cycle

Ethnography. The ethnographic research method, derived largely from the field of anthropology, emphasizes studying a phenomenon within the context of its culture. The researcher must be deeply immersed in the social culture over an extended period of time (usually 8 months to 2 years) and should engage, observe, and record the daily life of the studied culture and its social participants within their natural setting. The primary mode of data collection is participant observation, and data analysis involves a "sense-making" approach. In addition, the researcher must take extensive field notes, and narrate her experience in descriptive detail so that readers may experience the same culture as the researcher. In this method, the researcher has two roles: rely on her unique knowledge and engagement to generate insights (theory), and convince the scientific community of the trans-situational nature of the studied phenomenon.

The classic example of ethnographic research is Jane Goodall's study of primate behaviors, where she lived with chimpanzees in their natural habitat at Gombe National Park in Tanzania, observed their behaviors, interacted with them, and shared their lives. During that process, she learnt and chronicled how chimpanzees seek food and shelter, how they socialize with each other, their communication patterns, their mating behaviors, and so forth. A more contemporary example of ethnographic research is Myra Bluebond-Langer's (1996)[14] study of decision making in families with children suffering from life-threatening illnesses, and the physical, psychological, environmental, ethical, legal, and cultural issues that influence such decision-making. The researcher followed the experiences of approximately 80 children with

---

[14] Bluebond-Langer, M. (1996). *In the Shadow of Illness: Parents and Siblings of the Chronically Ill Child.* Princeton, NJ: Princeton University Press.

incurable illnesses and their families for a period of over two years. Data collection involved participant observation and formal/informal conversations with children, their parents and relatives, and health care providers to document their lived experience.

Phenomenology. Phenomenology is a research method that emphasizes the study of conscious experiences as a way of understanding the reality around us. It is based on the ideas of German philosopher Edmund Husserl in the early 20th century who believed that human experience is the source of all knowledge. Phenomenology is concerned with the systematic reflection and analysis of phenomena associated with conscious experiences, such as human judgment, perceptions, and actions, with the goal of (1) appreciating and describing social reality from the diverse subjective perspectives of the participants involved, and (2) understanding the symbolic meanings ("deep structure") underlying these subjective experiences. Phenomenological inquiry requires that researchers eliminate any prior assumptions and personal biases, empathize with the participant's situation, and tune into existential dimensions of that situation, so that they can fully understand the deep structures that drives the conscious thinking, feeling, and behavior of the studied participants.

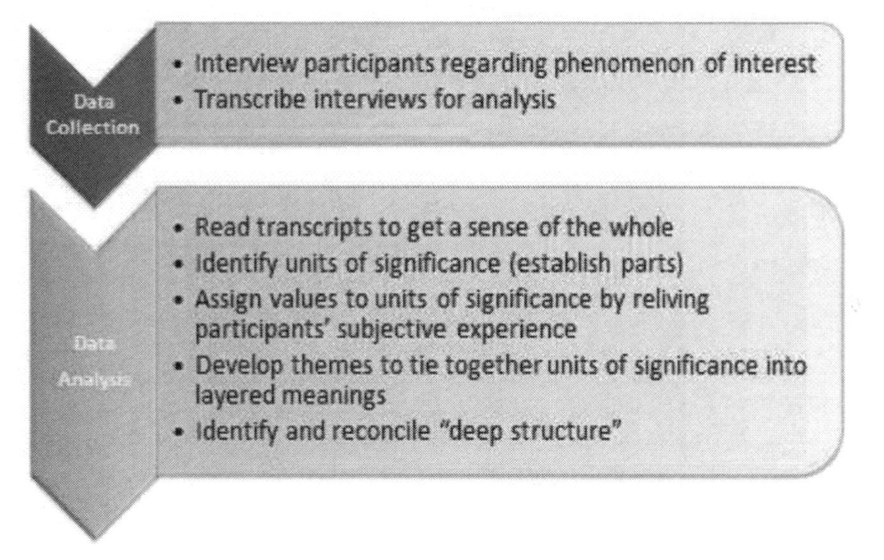

Figure 10.2. The existential phenomenological research method

Some researchers view phenomenology as a philosophy rather than as a research method. In response to this criticism, Giorgi and Giorgi (2003)[15] developed an *existential phenomenological research method* to guide studies in this area. This method, illustrated in Figure 10.2, can be grouped into data collection and data analysis phases. In the *data collection* phase, participants embedded in a social phenomenon are interviewed to capture their subjective experiences and perspectives regarding the phenomenon under investigation. Examples of questions that may be asked include "can you describe a typical day" or "can you describe that particular incident in more detail?" These interviews are recorded and transcribed for further analysis. During *data analysis*, the researcher reads the transcripts to: (1) get a sense of the whole, and (2) establish "units of significance" that can faithfully represent participants' subjective experiences. Examples of such units of significance are concepts such as "felt space" and "felt time," which are then used to document participants' psychological

---

[15] Giorgi, A and Giorgi, B (2003) Phenomenology. In J A Smith (ed.) *Qualitative Psychology: A Practical Guide to Research Methods*. London: Sage Publications.

experiences. For instance, did participants feel safe, free, trapped, or joyous when experiencing a phenomenon ("felt-space")? Did they feel that their experience was pressured, slow, or discontinuous ("felt-time")? Phenomenological analysis should take into account the participants' temporal landscape (i.e., their sense of past, present, and future), and the researcher must transpose herself in an imaginary sense in the participant's situation (i.e., temporarily live the participant's life). The participants' lived experience is described in form of a narrative or using emergent themes. The analysis then delves into these themes to identify multiple layers of meaning while retaining the fragility and ambiguity of subjects' lived experiences.

Rigor in Interpretive Research

While positivist research employs a "reductionist" approach by simplifying social reality into parsimonious theories and laws, interpretive research attempts to interpret social reality through the subjective viewpoints of the embedded participants within the context where the reality is situated. These interpretations are heavily contextualized, and are naturally less generalizable to other contexts. However, because interpretive analysis is subjective and sensitive to the experiences and insight of the embedded researcher, it is often considered less rigorous by many positivist (functionalist) researchers. Because interpretive research is based on different set of ontological and epistemological assumptions about social phenomenon than positivist research, the positivist notions of rigor, such as reliability, internal validity, and generalizability, do not apply in a similar manner. However, Lincoln and Guba (1985)[16] provide an alternative set of criteria that can be used to judge the rigor of interpretive research.

Dependability. Interpretive research can be viewed as dependable or authentic if two researchers assessing the same phenomenon using the same set of evidence independently arrive at the same conclusions or the same researcher observing the same or a similar phenomenon at different times arrives at similar conclusions. This concept is similar to that of reliability in positivist research, with agreement between two independent researchers being similar to the notion of inter-rater reliability, and agreement between two observations of the same phenomenon by the same researcher akin to test-retest reliability. To ensure dependability, interpretive researchers must provide adequate details about their phenomenon of interest and the social context in which it is embedded so as to allow readers to independently authenticate their interpretive inferences.

Credibility. Interpretive research can be considered credible if readers find its inferences to be believable. This concept is akin to that of *internal validity* in functionalistic research. The credibility of interpretive research can be improved by providing evidence of the researcher's extended engagement in the field, by demonstrating data triangulation across subjects or data collection techniques, and by maintaining meticulous data management and analytic procedures, such as verbatim transcription of interviews, accurate records of contacts and interviews, and clear notes on theoretical and methodological decisions, that can allow an independent audit of data collection and analysis if needed.

Confirmability. Confirmability refers to the extent to which the findings reported in interpretive research can be independently confirmed by others (typically, participants). This is similar to the notion of objectivity in functionalistic research. Since interpretive research rejects the notion of an objective reality, confirmability is demonstrated in terms of "inter-

---

[16] Lincoln, Y. S., and Guba, E. G. (1985). *Naturalistic Inquiry*. Beverly Hills, CA: Sage Publications.

subjectivity", i.e., if the study's participants agree with the inferences derived by the researcher. For instance, if a study's participants generally agree with the inferences drawn by a researcher about a phenomenon of interest (based on a review of the research paper or report), then the findings can be viewed as confirmable.

Transferability. Transferability in interpretive research refers to the extent to which the findings can be generalized to other settings. This idea is similar to that of external validity in functionalistic research. The researcher must provide rich, detailed descriptions of the research context ("thick description") and thoroughly describe the structures, assumptions, and processes revealed from the data so that readers can independently assess whether and to what extent are the reported findings transferable to other settings.

Chapter 13

Qualitative Analysis

Qualitative analysis is the analysis of qualitative data such as text data from interview transcripts. Unlike quantitative analysis, which is statistics driven and largely independent of the researcher, qualitative analysis is heavily dependent on the researcher's analytic and integrative skills and personal knowledge of the social context where the data is collected. The emphasis in qualitative analysis is "sense making" or understanding a phenomenon, rather than predicting or explaining. A creative and investigative mindset is needed for qualitative analysis, based on an ethically enlightened and participant-in-context attitude, and a set of analytic strategies. This chapter provides a brief overview of some of these qualitative analysis strategies. Interested readers are referred to more authoritative and detailed references such as Miles and Huberman's (1984)[17] seminal book on this topic.

Grounded Theory

How can you analyze a vast set qualitative data acquired through participant observation, in-depth interviews, focus groups, narratives of audio/video recordings, or secondary documents? One of these techniques for analyzing text data is **grounded theory** – an inductive technique of interpreting recorded data about a social phenomenon to build theories about that phenomenon. The technique was developed by Glaser and Strauss (1967)[18] in their *method of constant comparative analysis* of grounded theory research, and further refined by Strauss and Corbin (1990)[19] to further illustrate specific coding techniques – a process of classifying and categorizing text data segments into a set of codes (concepts), categories (constructs), and relationships. The interpretations are "grounded in" (or based on) observed empirical data, hence the name. To ensure that the theory is based solely on observed evidence, the grounded theory approach requires that researchers suspend any preexisting theoretical expectations or biases before data analysis, and let the data dictate the formulation of the theory.

Strauss and Corbin (1998) describe three coding techniques for analyzing text data: open, axial, and selective. **Open coding** is a process aimed at identifying **concepts** or key ideas

---

[17] Miles M. B., Huberman A. M. (1984). *Qualitative Data Analysis: A Sourcebook of New Methods*. Newbury Park, CA: Sage Publications.
[18] Glaser, B. and Strauss, A. (1967). *The Discovery of Grounded Theory: Strategies for Qualitative Research*, Chicago: Aldine.
[19] Strauss, A. and Corbin, J. (1990). *Basics of Qualitative Research: Grounded Theory Procedures and Techniques*, Beverly Hills, CA: Sage Publications.

that are hidden within textual data, which are potentially related to the phenomenon of interest. The researcher examines the raw textual data line by line to identify discrete events, incidents, ideas, actions, perceptions, and interactions of relevance that are coded as concepts (hence called *in vivo codes*). Each concept is linked to specific portions of the text (coding unit) for later validation. Some concepts may be simple, clear, and unambiguous while others may be complex, ambiguous, and viewed differently by different participants. The coding unit may vary with the concepts being extracted. Simple concepts such as "organizational size" may include just a few words of text, while complex ones such as "organizational mission" may span several pages. Concepts can be named using the researcher's own naming convention or standardized labels taken from the research literature. Once a basic set of concepts are identified, these concepts can then be used to code the remainder of the data, while simultaneously looking for new concepts and refining old concepts. While coding, it is important to identify the recognizable characteristics of each concept, such as its size, color, or level (e.g., high or low), so that similar concepts can be grouped together later. This coding technique is called "open" because the researcher is open to and actively seeking new concepts relevant to the phenomenon of interest.

Next, similar concepts are grouped into higher order **categories**. While concepts may be context-specific, categories tend to be broad and generalizable, and ultimately evolve into *constructs* in a grounded theory. Categories are needed to reduce the amount of concepts the researcher must work with and to build a "big picture" of the issues salient to understanding a social phenomenon. Categorization can be done is phases, by combining concepts into subcategories, and then subcategories into higher order categories. Constructs from the existing literature can be used to name these categories, particularly if the goal of the research is to extend current theories. However, caution must be taken while using existing constructs, as such constructs may bring with them commonly held beliefs and biases. For each category, its characteristics (or properties) and dimensions of each characteristic should be identified. The *dimension* represents a value of a characteristic along a continuum. For example, a "communication media" category may have a characteristic called "speed", which can be dimensionalized as fast, medium, or slow. Such categorization helps differentiate between different kinds of communication media and enables researchers identify *patterns* in the data, such as which communication media is used for which types of tasks.

The second phase of grounded theory is **axial coding**, where the categories and subcategories are assembled into causal relationships or hypotheses that can tentatively explain the phenomenon of interest. Although distinct from open coding, axial coding can be performed simultaneously with open coding. The relationships between categories may be clearly evident in the data or may be more subtle and implicit. In the latter instance, researchers may use a coding scheme (often called a "coding paradigm", but different from the paradigms discussed in Chapter 3) to understand which categories represent *conditions* (the circumstances in which the phenomenon is embedded), *actions/interactions* (the responses of individuals to events under these conditions), and *consequences* (the outcomes of actions/ interactions). As conditions, actions/interactions, and consequences are identified, theoretical propositions start to emerge, and researchers can start explaining why a phenomenon occurs, under what conditions, and with what consequences.

The third and final phase of grounded theory is **selective coding**, which involves identifying a central category or a core variable and systematically and logically relating this central category to other categories. The central category can evolve from existing categories or can be a higher order category that subsumes previously coded categories. New data is selectively sampled to validate the central category and its relationships to other categories (i.e., the tentative theory). Selective coding limits the range of analysis, and makes it move fast. At the same time, the coder must watch out for other categories that may emerge from the new data that may be related to the phenomenon of interest (open coding), which may lead to further refinement of the initial theory. Hence, open, axial, and selective coding may proceed simultaneously. Coding of new data and theory refinement continues until **theoretical saturation** is reached, i.e., when additional data does not yield any marginal change in the core categories or the relationships.

The "constant comparison" process implies continuous rearrangement, aggregation, and refinement of categories, relationships, and interpretations based on increasing depth of understanding, and an iterative interplay of four stages of activities: (1) comparing incidents/texts assigned to each category (to validate the category), (2) integrating categories and their properties, (3) delimiting the theory (focusing on the core concepts and ignoring less relevant concepts), and (4) writing theory (using techniques like memoing, storylining, and diagramming that are discussed in the next chapter). Having a central category does not necessarily mean that all other categories can be integrated nicely around it. In order to identify key categories that are conditions, action/interactions, and consequences of the core category, Strauss and Corbin (1990) recommend several integration techniques, such as storylining, memoing, or concept mapping. In **storylining**, categories and relationships are used to explicate and/or refine a story of the observed phenomenon. Memos are theorized write-ups of ideas about substantive concepts and their theoretically coded relationships as they evolve during ground theory analysis, and are important tools to keep track of and refine ideas that develop during the analysis. **Memoing** is the process of using these memos to discover patterns and relationships between categories using two-by-two tables, diagrams, or figures, or other illustrative displays. **Concept mapping** is a graphical representation of concepts and relationships between those concepts (e.g., using boxes and arrows). The major concepts are typically laid out on one or more sheets of paper, blackboards, or using graphical software programs, linked to each other using arrows, and readjusted to best fit the observed data.

After a grounded theory is generated, it must be refined for internal consistency and logic. Researchers must ensure that the central construct has the stated characteristics and dimensions, and if not, the data analysis may be repeated. Researcher must then ensure that the characteristics and dimensions of all categories show variation. For example, if behavior frequency is one such category, then the data must provide evidence of both frequent performers and infrequent performers of the focal behavior. Finally, the theory must be validated by comparing it with raw data. If the theory contradicts with observed evidence, the coding process may be repeated to reconcile such contradictions or unexplained variations.

Content Analysis

Content analysis is the systematic analysis of the content of a text (e.g., who says what, to whom, why, and to what extent and with what effect) in a quantitative or qualitative manner. Content analysis typically conducted as follows. First, when there are many texts to analyze (e.g., newspaper stories, financial reports, blog postings, online reviews, etc.), the researcher begins by sampling a selected set of texts from the population of texts for analysis. This process is not random, but instead, texts that have more pertinent content should be chosen selectively. Second, the researcher identifies and applies rules to divide each text into segments or "chunks" that can be treated as separate units of analysis. This process is called *unitizing*. For example, assumptions, effects, enablers, and barriers in texts may constitute such units. Third, the researcher constructs and applies one or more concepts to each unitized text segment in a process called *coding*. For coding purposes, a coding scheme is used based on the themes the researcher is searching for or uncovers as she classifies the text. Finally, the coded data is analyzed, often both quantitatively and qualitatively, to determine which themes occur most frequently, in what contexts, and how they are related to each other.

A simple type of content analysis is **sentiment analysis** – a technique used to capture people's opinion or attitude toward an object, person, or phenomenon. Reading online messages about a political candidate posted on an online forum and classifying each message as positive, negative, or neutral is an example of such an analysis. In this case, each message represents one unit of analysis. This analysis will help identify whether the sample as a whole is positively or negatively disposed or neutral towards that candidate. Examining the content of online reviews in a similar manner is another example. Though this analysis can be done manually, for very large data sets (millions of text records), natural language processing and text analytics based software programs are available to automate the coding process, and maintain a record of how people sentiments fluctuate with time.

A frequent criticism of content analysis is that it lacks a set of systematic procedures that would allow the analysis to be replicated by other researchers. Schilling (2006)[20] addressed this criticism by organizing different content analytic procedures into a spiral model. This model consists of five levels or phases in interpreting text: (1) convert recorded tapes into raw text data or transcripts for content analysis, (2) convert raw data into condensed protocols, (3) convert condensed protocols into a preliminary category system, (4) use the preliminary category system to generate coded protocols, and (5) analyze coded protocols to generate interpretations about the phenomenon of interest.

Content analysis has several limitations. First, the coding process is restricted to the information available in text form. For instance, if a researcher is interested in studying people's views on capital punishment, but no such archive of text documents is available, then the analysis cannot be done. Second, sampling must be done carefully to avoid sampling bias. For instance, if your population is the published research literature on a given topic, then you have systematically omitted unpublished research or the most recent work that is yet to be published.

Hermeneutic Analysis

Hermeneutic analysis is a special type of content analysis where the researcher tries to "interpret" the subjective meaning of a given text within its socio-historic context. Unlike grounded theory or content analysis, which ignores the context and meaning of text documents during the coding process, hermeneutic analysis is a truly interpretive technique for analyzing qualitative data. This method assumes that written texts narrate an author's experience within a socio-historic context, and should be interpreted as such within that context. Therefore, the researcher continually iterates between singular interpretation of the text (the part) and a holistic understanding of the context (the whole) to develop a fuller understanding of the phenomenon in its situated context, which German philosopher Martin Heidegger called the

[20] Schilling, J. (2006). "On the Pragmatics of Qualitative Assessment: Designing the Process for Content Analysis," *European Journal of Psychological Assessment* (22:1), 28-37.

hermeneutic circle. The word *hermeneutic* (singular) refers to one particular method or strand of interpretation.

More generally, hermeneutics is the study of interpretation and the theory and practice of interpretation. Derived from religious studies and linguistics, traditional hermeneutics, such as *biblical hermeneutics*, refers to the interpretation of written texts, especially in the areas of literature, religion and law (such as the Bible). In the 20th century, Heidegger suggested that a more direct, non-mediated, and authentic way of understanding social reality is to experience it, rather than simply observe it, and proposed *philosophical hermeneutics*, where the focus shifted from interpretation to existential understanding. Heidegger argued that texts are the means by which readers can not only read about an author's experience, but also relive the author's experiences. Contemporary or modern hermeneutics, developed by Heidegger's students such as Hans-Georg Gadamer, further examined the limits of written texts for communicating social experiences, and went on to propose a framework of the interpretive process, encompassing all forms of communication, including written, verbal, and non-verbal, and exploring issues that restrict the communicative ability of written texts, such as presuppositions, language structures (e.g., grammar, syntax, etc.), and semiotics (the study of written signs such as symbolism, metaphor, analogy, and sarcasm). The term hermeneutics is sometimes used interchangeably and inaccurately with *exegesis*, which refers to the interpretation or critical explanation of written text only and especially religious texts.

Conclusions

Finally, standard software programs, such as ATLAS.ti.5, NVivo, and QDA Miner, can be used to automate coding processes in qualitative research methods. These programs can quickly and efficiently organize, search, sort, and process large volumes of text data using user-defined rules. To guide such automated analysis, a coding schema should be created, specifying the keywords or codes to search for in the text, based on an initial manual examination of sample text data. The schema can be organized in a hierarchical manner to organize codes into higher-order codes or constructs. The coding schema should be validated using a different sample of texts for accuracy and adequacy. However, if the coding schema is biased or incorrect, the resulting analysis of the entire population of text may be flawed and non-interpretable. However, software programs cannot decipher the meaning behind the certain words or phrases or the context within which these words or phrases are used (such as those in sarcasms or metaphors), which may lead to significant misinterpretation in large scale qualitative analysis.

Chapter 14

Quantitative Analysis: Descriptive Statistics

Numeric data collected in a research project can be analyzed quantitatively using statistical tools in two different ways. **Descriptive analysis** refers to statistically describing, aggregating, and presenting the constructs of interest or associations between these constructs. **Inferential analysis** refers to the statistical testing of hypotheses (theory testing). In this chapter, we will examine statistical techniques used for descriptive analysis, and the next chapter will examine statistical techniques for inferential analysis. Much of today's quantitative data analysis is conducted using software programs such as SPSS or SAS. Readers are advised to familiarize themselves with one of these programs for understanding the concepts described in this chapter.

Data Preparation

In research projects, data may be collected from a variety of sources: mail-in surveys, interviews, pretest or posttest experimental data, observational data, and so forth. This data must be converted into a machine-readable, numeric format, such as in a spreadsheet or a text file, so that they can be analyzed by computer programs like SPSS or SAS. Data preparation usually follows the following steps.

Data coding. Coding is the process of converting data into numeric format. A codebook should be created to guide the coding process. A **codebook** is a comprehensive document containing detailed description of each variable in a research study, items or measures for that variable, the format of each item (numeric, text, etc.), the response scale for each item (i.e., whether it is measured on a nominal, ordinal, interval, or ratio scale; whether such scale is a five-point, seven-point, or some other type of scale), and how to code each value into a numeric format. For instance, if we have a measurement item on a seven-point Likert scale with anchors ranging from "strongly disagree" to "strongly agree", we may code that item as 1 for strongly disagree, 4 for neutral, and 7 for strongly agree, with the intermediate anchors in between. Nominal data such as industry type can be coded in numeric form using a coding scheme such as: 1 for manufacturing, 2 for retailing, 3 for financial, 4 for healthcare, and so forth (of course, nominal data cannot be analyzed statistically). Ratio scale data such as age, income, or test scores can be coded as entered by the respondent. Sometimes, data may need to be aggregated into a different form than the format used for data collection. For instance, for measuring a construct such as "benefits of computers," if a survey provided respondents with a checklist of benefits that they could select from (i.e., they could choose as many of those benefits as they wanted), then the total number of checked items can be used as an aggregate measure of benefits. Note that many other forms of data, such as interview transcripts, cannot be converted into a numeric format for statistical analysis. Coding is especially important for large complex studies involving many variables and measurement items, where the coding process is conducted by different people, to help the coding team code data in a consistent manner, and also to help others understand and interpret the coded data.

Data entry. Coded data can be entered into a spreadsheet, database, text file, or directly into a statistical program like SPSS. Most statistical programs provide a data editor for entering data. However, these programs store data in their own native format (e.g., SPSS stores data as .sav files), which makes it difficult to share that data with other statistical programs. Hence, it is often better to enter data into a spreadsheet or database, where they can be reorganized as needed, shared across programs, and subsets of data can be extracted for analysis. Smaller data sets with less than 65,000 observations and 256 items can be stored in a spreadsheet such as Microsoft Excel, while larger dataset with millions of observations will require a database. Each observation can be entered as one row in the spreadsheet and each measurement item can be represented as one column. The entered data should be frequently checked for accuracy, via occasional spot checks on a set of items or observations, during and after entry. Furthermore, while entering data, the coder should watch out for obvious evidence of bad data, such as the respondent selecting the "strongly agree" response to all items irrespective of content, including reverse-coded items. If so, such data can be entered but should be excluded from subsequent analysis.

Missing values. Missing data is an inevitable part of any empirical data set. Respondents may not answer certain questions if they are ambiguously worded or too sensitive. Such problems should be detected earlier during pretests and corrected before the main data collection process begins. During data entry, some statistical programs automatically treat blank entries as missing values, while others require a specific numeric value such as -1 or 999 to be entered to denote a missing value. During data analysis, the default mode of handling missing values in most software programs is to simply drop the entire observation containing even a single missing value, in a technique called *listwise deletion*. Such deletion can significantly shrink the sample size and make it extremely difficult to detect small effects. Hence, some software programs allow the option of replacing missing values with an estimated value via a process called *imputation*. For instance, if the missing value is one item in a multi-item scale, the imputed value may be the average of the respondent's responses to remaining items on that scale. If the missing value belongs to a single-item scale, many researchers use the average of other respondent's responses to that item as the imputed value. Such imputation may be biased if the missing value is of a systematic nature rather than a random nature. Two methods that can produce relatively unbiased estimates for imputation are the maximum likelihood procedures and multiple imputation methods, both of which are supported in popular software programs such as SPSS and SAS.

Data transformation. Sometimes, it is necessary to transform data values before they can be meaningfully interpreted. For instance, reverse coded items, where items convey the opposite meaning of that of their underlying construct, should be reversed (e.g., in a 1-7 interval scale, 8 minus the observed value will reverse the value) before they can be compared or combined with items that are not reverse coded. Other kinds of transformations may include creating scale measures by adding individual scale items, creating a weighted index from a set of observed measures, and collapsing multiple values into fewer categories (e.g., collapsing incomes into income ranges).

Univariate Analysis

Univariate analysis, or analysis of a single variable, refers to a set of statistical techniques that can describe the general properties of one variable. Univariate statistics include: (1) frequency distribution, (2) central tendency, and (3) dispersion. The **frequency distribution** of a variable is a summary of the frequency (or percentages) of individual values or ranges of values for that variable. For instance, we can measure how many times a sample of respondents attend religious services (as a measure of their "religiosity") using a categorical scale: never, once per year, several times per year, about once a month, several times per month, several times per week, and an optional category for "did not answer." If we count the number (or percentage) of observations within each category (except "did not answer" which is really a missing value rather than a category), and display it in the form of a table as shown in Figure 14.1, what we have is a frequency distribution. This distribution can also be depicted in the form of a bar chart, as shown on the right panel of Figure 14.1, with the horizontal axis representing each category of that variable and the vertical axis representing the frequency or percentage of observations within each category.

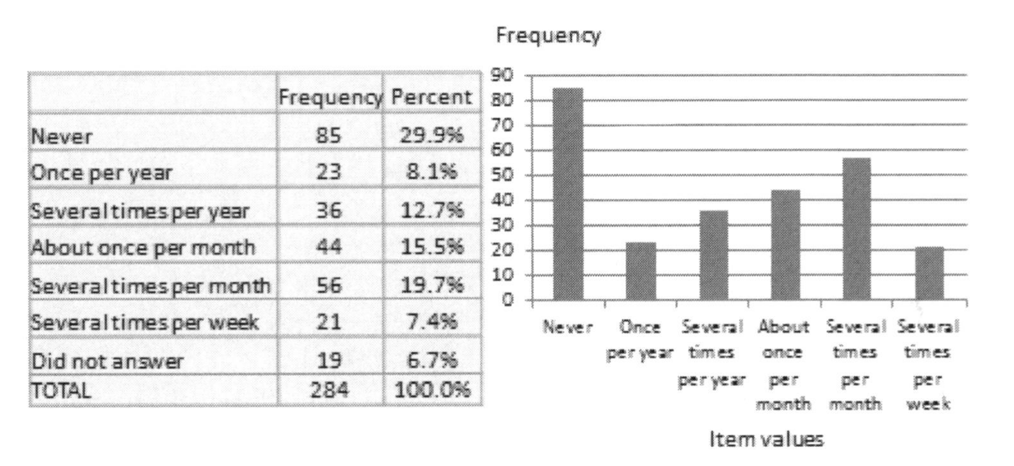

	Frequency	Percent
Never	85	29.9%
Once per year	23	8.1%
Several times per year	36	12.7%
About once per month	44	15.5%
Several times per month	56	19.7%
Several times per week	21	7.4%
Did not answer	19	6.7%
TOTAL	284	100.0%

Figure 14.1. Frequency distribution of religiosity

With very large samples where observations are independent and random, the frequency distribution tends to follow a plot that looked like a bell-shaped curve (a smoothed bar chart of the frequency distribution) similar to that shown in Figure 14.2, where most observations are clustered toward the center of the range of values, and fewer and fewer observations toward the extreme ends of the range. Such a curve is called a *normal distribution*.

Central tendency is an estimate of the center of a distribution of values. There are three major estimates of central tendency: mean, median, and mode. The **arithmetic mean** (often simply called the "mean") is the simple average of all values in a given distribution. Consider a set of eight test scores: 15, 22, 21, 18, 36, 15, 25, 15. The arithmetic mean of these values is (15 + 20 + 21 + 20 + 36 + 15 + 25 + 15)/8 = 20.875. Other types of means include *geometric mean* (n^{th} root of the product of n numbers in a distribution) and *harmonic mean* (the reciprocal of the arithmetic means of the reciprocal of each value in a distribution), but these means are not very popular for statistical analysis of social research data.

The second measure of central tendency, the **median**, is the middle value within a range of values in a distribution. This is computed by sorting all values in a distribution in increasing order and selecting the middle value. In case there are two middle values (if there is an even number of values in a distribution), the average of the two middle values represent the median. In the above example, the sorted values are: 15, 15, 15, 18, 22, 21, 25, 36. The two middle values are 18 and 22, and hence the median is (18 + 22)/2 = 20.

Lastly, the **mode** is the most frequently occurring value in a distribution of values. In the previous example, the most frequently occurring value is 15, which is the mode of the above set of test scores. Note that any value that is estimated from a sample, such as mean, median, mode, or any of the later estimates are called a **statistic**.

Dispersion refers to the way values are *spread* around the central tendency, for example, how tightly or how widely are the values clustered around the mean. Two common measures of dispersion are the range and standard deviation. The **range** is the difference between the highest and lowest values in a distribution. The range in our previous example is 36-15 = 21.

The range is particularly sensitive to the presence of outliers. For instance, if the highest value in the above distribution was 85 and the other vales remained the same, the range would be 85-15 = 70. **Standard deviation**, the second measure of dispersion, corrects for such outliers by using a formula that takes into account how close or how far each value from the distribution mean:

$$\sigma = \sqrt{\frac{\sum_{i=1}^{n} (x_i - \mu)^2}{n - 1}}$$

where σ is the standard deviation, x_i is the ith observation (or value), μ is the arithmetic mean, n is the total number of observations, and Σ means summation across all observations. The square of the standard deviation is called the **variance** of a distribution. In a normally distributed frequency distribution, it is seen that 68% of the observations lie within one standard deviation of the mean ($\mu \pm 1\ \sigma$), 95% of the observations lie within two standard deviations ($\mu \pm 2\ \sigma$), and 99.7% of the observations lie within three standard deviations ($\mu \pm 3\ \sigma$), as shown in Figure 14.2.

Bivariate Analysis

Bivariate analysis examines how two variables are related to each other. The most common bivariate statistic is the **bivariate correlation** (often, simply called "correlation"), which is a number between -1 and +1 denoting the strength of the relationship between two variables. Let's say that we wish to study how age is related to self-esteem in a sample of 20 respondents, i.e., as age increases, does self-esteem increase, decrease, or remains unchanged. If self-esteem increases, then we have a positive correlation between the two variables, if self-esteem decreases, we have a negative correlation, and if it remains the same, we have a zero correlation. To calculate the value of this correlation, consider the hypothetical dataset shown in Table 14.1.

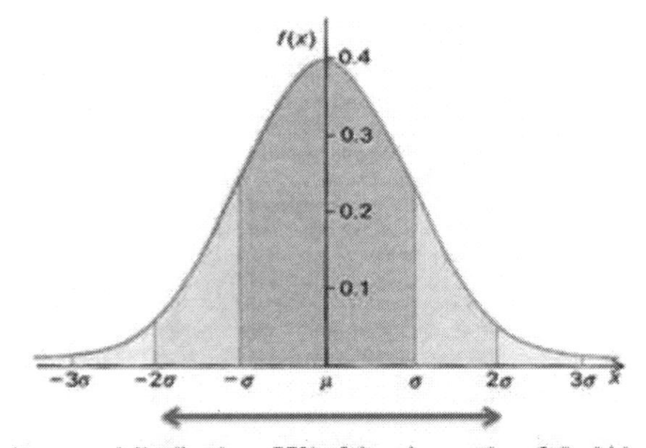

In normal distributions, 95% of the observations fall within two standard deviations of the mean value (μ ± 2σ)

Figure 14.2. Normal distribution

Obs	Age (x)	Self-Esteem (y)	xy	x^2	y^2
1	39	4.1	159.9	1521	16.81
2	45	4.6	207	2025	21.16
3	29	3.8	110.2	841	14.44
4	42	4.4	184.8	1764	19.36
5	19	3.2	60.8	361	10.24
6	22	3.1	68.2	484	9.61
7	39	3.8	148.2	1521	14.44
8	30	4.1	123	900	16.81
9	33	4.3	141.9	1089	18.49
10	23	3.7	85.1	529	13.69
11	20	3.5	70	400	12.25
12	18	3.2	57.6	324	10.24
13	24	3.7	88.8	576	13.69
14	22	3.3	72.6	484	10.89
15	29	3.4	98.6	841	11.56
16	35	4.0	140	1225	16.00
17	36	4.1	147.6	1296	16.81
18	37	3.8	140.6	1369	14.44
19	35	3.4	119	1225	11.56
20	32	3.6	115.2	1024	12.96
Sum (Σ)	609	75.1	2339.1	19799.0	285.45

Table 14.1. Hypothetical data on age and self-esteem

The two variables in this dataset are age (x) and self-esteem (y). Age is a ratio-scale variable, while self-esteem is an average score computed from a multi-item self-esteem scale measured using a 7-point Likert scale, ranging from "strongly disagree" to "strongly agree." The histogram of each variable is shown on the left side of Figure 14.3. The formula for calculating bivariate correlation is:

$$r_{xy} = \frac{\sum x_i y_i - n\bar{x}\bar{y}}{(n-1)s_x s_y} = \frac{n\sum x_i y_i - \sum x_i \sum y_i}{\sqrt{n\sum x_i^2 - (\sum x_i)^2}\ \sqrt{n\sum y_i^2 - (\sum y_i)^2}}.$$

where r_{xy} is the correlation, x and y are the sample means of x and y, and s_x and s_y are the standard deviations of x and y. The manually computed value of correlation between age and self-esteem, using the above formula as shown in Table 14.1, is 0.79. This figure indicates that age has a strong positive correlation with self-esteem, i.e., self-esteem tends to increase with increasing age, and decrease with decreasing age. Such pattern can also be seen from visually comparing the age and self-esteem histograms shown in Figure 14.3, where it appears that the top of the two histograms generally follow each other. Note here that the vertical axes in Figure 14.3 represent actual observation values, and not the frequency of observations (as was in Figure 14.1), and hence, these are not frequency distributions but rather histograms. The bivariate scatter plot in the right panel of Figure 14.3 is essentially a plot of self-esteem on the vertical axis against age on the horizontal axis. This plot roughly resembles an upward sloping line (i.e., positive slope), which is also indicative of a positive correlation. If the two variables were negatively correlated, the scatter plot would slope down (negative slope), implying that an increase in age would be related to a decrease in self-esteem and vice versa. If the two variables were uncorrelated, the scatter plot would approximate a horizontal line (zero slope), implying than an increase in age would have no systematic bearing on self-esteem.

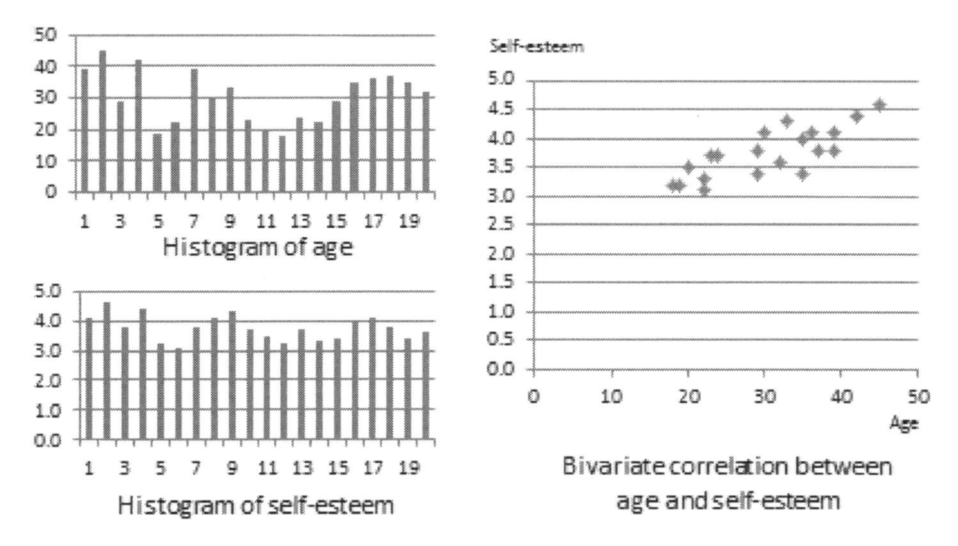

Figure 14.3. Histogram and correlation plot of age and self-esteem

After computing bivariate correlation, researchers are often interested in knowing whether the correlation is significant (i.e., a real one) or caused by mere chance. Answering such a question would require testing the following hypothesis:

$$H_0: r = 0$$
$$H_1: r \neq 0$$

H_0 is called the *null hypotheses*, and H_1 is called the *alternative hypothesis* (sometimes, also represented as H_a). Although they may seem like two hypotheses, H_0 and H_1 actually represent a single hypothesis since they are direct opposites of each other. We are interested in testing H_1 rather than H_0. Also note that H_1 is a non-directional hypotheses since it does not specify whether r is greater than or less than zero. Directional hypotheses will be specified as H_0: $r \leq 0$; H_1: $r > 0$ (if we are testing for a positive correlation). Significance testing of directional hypothesis is done using a one-tailed t-test, while that for non-directional hypothesis is done using a two-tailed t-test.

In statistical testing, the alternative hypothesis cannot be tested directly. Rather, it is tested indirectly by rejecting the null hypotheses with a certain level of probability. Statistical testing is always probabilistic, because we are never sure if our inferences, based on sample data, apply to the population, since our sample never equals the population. The probability that a statistical inference is caused pure chance is called the **p-value**. The p-value is compared with the **significance level** (α), which represents the maximum level of risk that we are willing to take that our inference is incorrect. For most statistical analysis, α is set to 0.05. A p-value less than $\alpha=0.05$ indicates that we have enough statistical evidence to reject the null hypothesis, and thereby, indirectly accept the alternative hypothesis. If $p>0.05$, then we do not have adequate statistical evidence to reject the null hypothesis or accept the alternative hypothesis.

The easiest way to test for the above hypothesis is to look up critical values of r from statistical tables available in any standard text book on statistics or on the Internet (most software programs also perform significance testing). The critical value of r depends on our desired significance level ($\alpha = 0.05$), the degrees of freedom (df), and whether the desired test is a one-tailed or two-tailed test. The **degree of freedom** is the number of values that can vary freely in any calculation of a statistic. In case of correlation, the df simply equals n – 2, or for the data in Table 14.1, df is 20 – 2 = 18. There are two different statistical tables for one-tailed and two-tailed test. In the two-tailed table, the critical value of r for $\alpha = 0.05$ and df = 18 is 0.44. For our computed correlation of 0.79 to be significant, it must be larger than the critical value of 0.44 or less than -0.44. Since our computed value of 0.79 is greater than 0.44, we conclude that there is a significant correlation between age and self-esteem in our data set, or in other words, the odds are less than 5% that this correlation is a chance occurrence. Therefore, we can reject the null hypotheses that $r \leq 0$, which is an indirect way of saying that the alternative hypothesis $r > 0$ is probably correct.

Most research studies involve more than two variables. If there are n variables, then we will have a total of n*(n-1)/2 possible correlations between these n variables. Such correlations are easily computed using a software program like SPSS, rather than manually using the formula for correlation (as we did in Table 14.1), and represented using a correlation matrix, as shown in Table 14.2. A correlation matrix is a matrix that lists the variable names along the first row and the first column, and depicts bivariate correlations between pairs of variables in the appropriate cell in the matrix. The values along the principal diagonal (from the top left to the bottom right corner) of this matrix are always 1, because any variable is always perfectly correlated with itself. Further, since correlations are non-directional, the correlation between variables V1 and V2 is the same as that between V2 and V1. Hence, the lower triangular matrix (values below the principal diagonal) is a mirror reflection of the upper triangular matrix (values above the principal diagonal), and therefore, we often list only the lower triangular matrix for simplicity. If the correlations involve variables measured using interval scales, then this specific type of correlations are called **Pearson product moment correlations**.

Another useful way of presenting bivariate data is cross-tabulation (often abbreviated to cross-tab, and sometimes called more formally as a contingency table). A **cross-tab** is a table that describes the frequency (or percentage) of all combinations of two or more nominal or categorical variables. As an example, let us assume that we have the following observations of gender and grade for a sample of 20 students, as shown in Figure 14.3. Gender is a nominal variable (male/female or M/F), and grade is a categorical variable with three levels (A, B, and C). A simple cross-tabulation of the data may display the joint distribution of gender and grades (i.e., how many students of each gender are in each grade category, as a raw frequency count or as a percentage) in a 2 x 3 matrix. This matrix will help us see if A, B, and C grades are equally distributed across male and female students. The cross-tab data in Table 14.3 shows that the distribution of A grades is biased heavily toward female students: in a sample of 10 male and 10 female students, five female students received the A grade compared to only one male students. In contrast, the distribution of C grades is biased toward male students: three male students received a C grade, compared to only one female student. However, the distribution of B grades was somewhat uniform, with six male students and five female students. The last row and the last column of this table are called marginal totals because they indicate the totals across each category and displayed along the margins of the table.

	V1	V2	V3	V4	V5	V6	V7	V8
V1	1.000							
V2	0.274	1.000						
V3	-0.134	-0.269	1.000					
V4	0.201	-0.153	0.075	1.000				
V5	-0.095	-0.166	0.278	-0.011	1.000			
V6	-0.129	0.280	-0.348	-0.378	-0.009	1.000		
V7	0.171	-0.122	0.296	0.086	0.193	0.233	1.000	
V8	0.518	0.238	0.238	-0.227	-0.551	0.082	-0.102	1.000

Table 14.2. A hypothetical correlation matrix for eight variables

Obs	1	2	3	4	5	6	7	8	9	10	11	12	13	14	15	16	17	18	19	20
Gender	F	M	F	M	F	M	M	M	F	F	M	M	M	F	F	M	F	F	F	M
Grade	A	B	B	B	C	A	C	B	B	A	B	C	C	B	B	B	A	A	B	B

Hypothetical data set

			Grades			Total
			A	B	C	
Gender	Male	Count	1	6	3	10
		Expected count	2.5	5.5	2.0	
	Female	Count	4	5	1	10
		Expected count	2.5	5.5	2.0	
Total			5	11	4	20

Cross-tabulation of gender versus age

Table 14.3. Example of cross-tab analysis

Although we can see a distinct pattern of grade distribution between male and female students in Table 14.3, is this pattern real or "statistically significant"? In other words, do the above frequency counts differ from that that may be expected from pure chance? To answer this question, we should compute the expected count of observation in each cell of the 2 x 3 cross-tab matrix. This is done by multiplying the marginal column total and the marginal row total for each cell and dividing it by the total number of observations. For example, for the male/A grade cell, expected count = 5 * 10 / 20 = 2.5. In other words, we were expecting 2.5 male students to receive an A grade, but in reality, only one student received the A grade. Whether this difference between expected and actual count is significant can be tested using a *chi-square test*. The chi-square statistic can be computed as the average difference between observed and expected counts across all cells. We can then compare this number to the critical value associated with a desired probability level ($p < 0.05$) and the degrees of freedom, which is simply (m-1)*(n-1), where m and n are the number of rows and columns respectively. In this example, df = (2 – 1) * (3 – 1) = 2. From standard chi-square tables in any statistics book, the critical chi-square value for p=0.05 and df=2 is 5.99. The computed chi-square value, based on our observed data, is 1.00, which is less than the critical value. Hence, we must conclude that the observed grade pattern is not statistically different from the pattern that can be expected by pure chance.

Chapter 15

Quantitative Analysis: Inferential Statistics

Inferential statistics are the statistical procedures that are used to reach conclusions about associations between variables. They differ from descriptive statistics in that they are explicitly designed to test hypotheses. Numerous statistical procedures fall in this category, most of which are supported by modern statistical software such as SPSS and SAS. This chapter provides a short primer on only the most basic and frequent procedures; readers are advised to consult a formal text on statistics or take a course on statistics for more advanced procedures.

Basic Concepts

British philosopher Karl Popper said that theories can never be proven, only disproven. As an example, how can we prove that the sun will rise tomorrow? Popper said that just because the sun has risen every single day that we can remember does not necessarily mean that it will rise tomorrow, because inductively derived theories are only conjectures that may or may not be predictive of future phenomenon. Instead, he suggested that we may assume a theory that the sun will rise every day without necessarily proving it, and if the sun does not rise on a certain day, the theory is falsified and rejected. Likewise, we can only reject hypotheses based on contrary evidence but can never truly accept them because presence of evidence does not mean that we may not observe contrary evidence later. Because we cannot truly accept a hypothesis of interest (alternative hypothesis), we formulate a null hypothesis as the opposite of the alternative hypothesis, and then use empirical evidence to reject the null hypothesis to demonstrate indirect, probabilistic support for our alternative hypothesis.

A second problem with testing hypothesized relationships in social science research is that the dependent variable may be influenced by an infinite number of extraneous variables and it is not plausible to measure and control for all of these extraneous effects. Hence, even if two variables may seem to be related in an observed sample, they may not be truly related in the population, and therefore inferential statistics are never certain or deterministic, but always probabilistic.

How do we know whether a relationship between two variables in an observed sample is significant, and not a matter of chance? Sir Ronald A. Fisher, one of the most prominent statisticians in history, established the basic guidelines for significance testing. He said that a statistical result may be considered significant if it can be shown that the probability of it being rejected due to chance is 5% or less. In inferential statistics, this probability is called the **p-value**, 5% is called the significance level (α), and the desired relationship between the p-value and α is denoted as: $p \leq 0.05$. The **significance level** is the maximum level of risk that we are willing to accept as the price of our inference from the sample to the population. If the p-value is less than 0.05 or 5%, it means that we have a 5% chance of being incorrect in rejecting the null hypothesis or having a Type I error. If $p > 0.05$, we do not have enough evidence to reject the null hypothesis or accept the alternative hypothesis.

We must also understand three related statistical concepts: sampling distribution, standard error, and confidence interval. A **sampling distribution** is the theoretical distribution of an infinite number of samples from the population of interest in your study. However, because a sample is never identical to the population, every sample always has some inherent level of error, called the **standard error**. If this standard error is small, then statistical estimates derived from the sample (such as sample mean) are reasonably good estimates of the population. The precision of our sample estimates is defined in terms of a **confidence interval** (CI). A 95% CI is defined as a range of plus or minus two standard deviations of the mean estimate, as derived from different samples in a sampling distribution. Hence, when we say that our observed sample estimate has a CI of 95%, what we mean is that we are confident that 95% of the time, the population parameter is within two standard deviations of our observed sample estimate. Jointly, the p-value and the CI give us a good idea of the probability of our result and how close it is from the corresponding population parameter.

General Linear Model

Most inferential statistical procedures in social science research are derived from a general family of statistical models called the **general linear model** (GLM). A *model* is an estimated mathematical equation that can be used to represent a set of data, and *linear* refers to a straight line. Hence, a GLM is a system of equations that can be used to represent linear patterns of relationships in observed data.

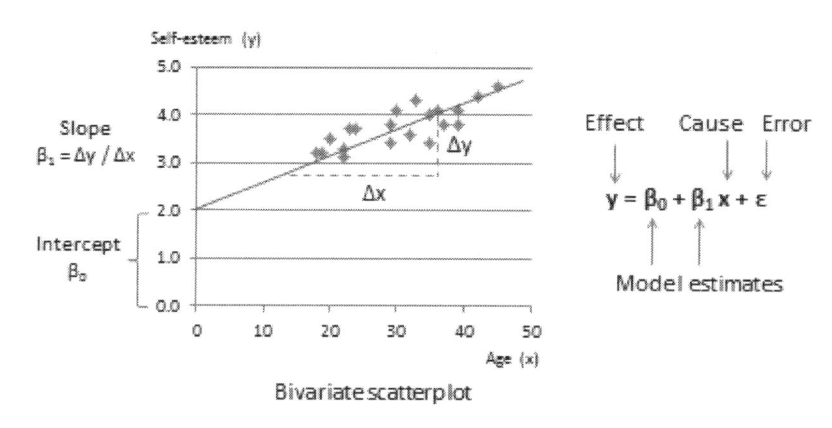

Figure 15.1. Two-variable linear model

The simplest type of GLM is a two-variable linear model that examines the relationship between one independent variable (the cause or predictor) and one dependent variable (the effect or outcome). Let us assume that these two variables are age and self-esteem respectively. The bivariate scatterplot for this relationship is shown in Figure 15.1, with age (predictor) along the horizontal or x-axis and self-esteem (outcome) along the vertical or y-axis. From the scatterplot, it appears that individual observations representing combinations of age and self-esteem generally seem to be scattered around an imaginary upward sloping straight line. We can estimate parameters of this line, such as its slope and intercept from the GLM. From high-school algebra, recall that straight lines can be represented using the mathematical equation y = mx + c, where m is the *slope* of the straight line (how much does y change for unit change in x) and c is the *intercept* term (what is the value of y when x is zero). In GLM, this equation is represented formally as:

$$y = \beta_0 + \beta_1 x + \varepsilon$$

where β_0 is the slope, β_1 is the intercept term, and ε is the *error term*. ε represents the deviation of actual observations from their estimated values, since most observations are close to the line but do not fall exactly on the line (i.e., the GLM is not perfect). Note that a linear model can have more than two predictors. To visualize a linear model with two predictors, imagine a three-dimensional cube, with the outcome (y) along the vertical axis, and the two predictors (say, x_1 and x_2) along the two horizontal axes along the base of the cube. A line that describes the relationship between two or more variables is called a regression line, β_0 and β_1 (and other beta values) are called *regression coefficients*, and the process of estimating regression coefficients is called **regression analysis**. The GLM for regression analysis with n predictor variables is:

$$y = \beta_0 + \beta_1 x_1 + \beta_2 x_2 + \beta_3 x_3 + \dots + \beta_n x_n + \varepsilon$$

In the above equation, predictor variables x_i may represent independent variables or *covariates* (control variables). Covariates are variables that are not of theoretical interest but may have some impact on the dependent variable y and should be controlled, so that the residual effects of the independent variables of interest are detected more precisely. Covariates capture systematic errors in a regression equation while the error term (ε) captures random errors. Though most variables in the GLM tend to be interval or ratio-scaled, this does not have to be the case. Some predictor variables may even be nominal variables (e.g., gender: male or female), which are coded as *dummy variables*. These are variables that can assume one of only two possible values: 0 or 1 (in the gender example, "male" may be designated as 0 and "female" as 1 or vice versa). A set of n nominal variables is represented using n–1 dummy variables. For instance, industry sector, consisting of the agriculture, manufacturing, and service sectors, may be represented using a combination of two dummy variables (x_1, x_2), with (0, 0) for agriculture, (0, 1) for manufacturing, and (1, 1) for service. It does not matter which level of a nominal variable is coded as 0 and which level as 1, because 0 and 1 values are treated as two distinct groups (such as treatment and control groups in an experimental design), rather than as numeric quantities, and the statistical parameters of each group are estimated separately.

The GLM is a very powerful statistical tool because it is not one single statistical method, but rather a family of methods that can be used to conduct sophisticated analysis with different types and quantities of predictor and outcome variables. If we have a dummy predictor variable, and we are comparing the effects of the two levels (0 and 1) of this dummy variable on the outcome variable, we are doing an *analysis of variance* (ANOVA). If we are doing ANOVA while controlling for the effects of one or more covariate, we have an *analysis of covariance* (ANCOVA). We can also have multiple outcome variables (e.g., y_1, y_1, ... y_n), which are represented using a "system of equations" consisting of a different equation for each outcome variable (each with its own unique set of regression coefficients). If multiple outcome variables are modeled as being predicted by the same set of predictor variables, the resulting analysis is called *multivariate regression*. If we are doing ANOVA or ANCOVA analysis with multiple outcome variables, the resulting analysis is a *multivariate ANOVA* (MANOVA) or *multivariate ANCOVA* (MANCOVA) respectively. If we model the outcome in one regression equation as a predictor in another equation in an interrelated system of regression equations, then we have a very sophisticated type of analysis called *structural equation modeling*. The most important problem in GLM is *model specification*, i.e., how to specify a regression equation (or a system of equations) to best represent the phenomenon of interest. Model specification should be based on theoretical considerations about the phenomenon being studied, rather than what fits the observed data best. The role of data is in validating the model, and not in its specification.

Two-Group Comparison

One of the simplest inferential analyses is comparing the post-test outcomes of treatment and control group subjects in a randomized post-test only control group design, such as whether students enrolled to a special program in mathematics perform better than those in a traditional math curriculum. In this case, the predictor variable is a dummy variable (1=treatment group, 0=control group), and the outcome variable, performance, is ratio scaled (e.g., score of a math test following the special program). The analytic technique for this simple design is a *one-way ANOVA* (one-way because it involves only one predictor variable), and the statistical test used is called a *Student's t-test* (or t-test, in short).

The t-test was introduced in 1908 by William Sealy Gosset, a chemist working for the Guiness Brewery in Dublin, Ireland to monitor the quality of stout – a dark beer popular with 19th century porters in London. Because his employer did not want to reveal the fact that it was using statistics for quality control, Gosset published the test in *Biometrika* using his pen name "Student" (he was a student of Sir Ronald Fisher), and the test involved calculating the value of t, which was a letter used frequently by Fisher to denote the difference between two groups. Hence, the name Student's t-test, although Student's identity was known to fellow statisticians.

The t-test examines whether the means of two groups are statistically different from each other (non-directional or two-tailed test), or whether one group has a statistically larger (or smaller) mean than the other (directional or one-tailed test). In our example, if we wish to examine whether students in the special math curriculum perform better than those in traditional curriculum, we have a one-tailed test. This hypothesis can be stated as:

$$H_0: \mu_1 \leq \mu_2 \qquad \text{(null hypothesis)}$$
$$H_1: \mu_1 > \mu_2 \qquad \text{(alternative hypothesis)}$$

where μ_1 represents the mean population performance of students exposed to the special curriculum (treatment group) and μ_2 is the mean population performance of students with traditional curriculum (control group). Note that the null hypothesis is always the one with the "equal" sign, and the goal of all statistical significance tests is to reject the null hypothesis.

How can we infer about the difference in population means using data from samples drawn from each population? From the hypothetical frequency distributions of the treatment and control group scores in Figure 15.2, the control group appears to have a bell-shaped (normal) distribution with a mean score of 45 (on a 0-100 scale), while the treatment group appear to have a mean score of 65. These means look different, but they are really sample means ($\bar{X}$), which may differ from their corresponding population means (μ) due to sampling error. Sample means are probabilistic estimates of population means within a certain confidence interval (95% CI is sample mean $\pm$ two standard errors, where standard error is the standard deviation of the distribution in sample means as taken from infinite samples of the population. Hence, statistical significance of population means depends not only on sample mean scores, but also on the standard error or the degree of spread in the frequency distribution of the sample means. If the spread is large (i.e., the two bell-shaped curves have a lot of overlap), then the 95% CI of the two means may also be overlapping, and we cannot conclude with high probability ($p<0.05$) that that their corresponding population means are significantly different. However, if the curves have narrower spreads (i.e., they are less overlapping), then the CI of each mean may not overlap, and we reject the null hypothesis and say that the population means of the two groups are significantly different at $p<0.05$.

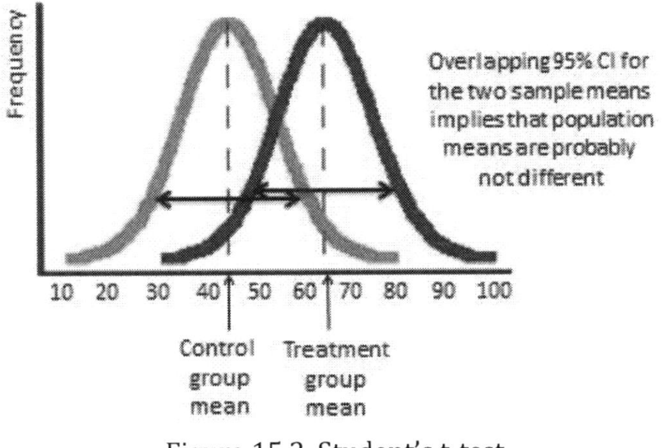

Figure 15.2. Student's t-test

To conduct the t-test, we must first compute a t-statistic of the difference is sample means between the two groups. This statistic is the ratio of the difference in sample means relative to the difference in their variability of scores (standard error):

$$t = \frac{\overline{X}_1 - \overline{X}_2}{s_{\overline{X}_1 - \overline{X}_2}}$$

where the numerator is the difference in sample means between the treatment group (Group 1) and the control group (Group 2) and the denominator is the standard error of the difference between the two groups, which in turn, can be estimated as:

$$s_{\overline{X}_1 - \overline{X}_2} = \sqrt{\frac{s_1^2}{n_1} + \frac{s_2^2}{n_2}}.$$

s^2 is the variance and n is the sample size of each group. The t-statistic will be positive if the treatment mean is greater than the control mean. To examine if this t-statistic is large enough than that possible by chance, we must look up the probability or p-value associated with our computed t-statistic in statistical tables available in standard statistics text books or on the Internet or as computed by statistical software programs such as SAS and SPSS. This value is a function of the t-statistic, whether the t-test is one-tailed or two-tailed, and the *degrees of freedom* (df) or the number of values that can vary freely in the calculation of the statistic (usually a function of the sample size and the type of test being performed). The degree of freedom of the t-statistic is computed as:

$$\text{d.f.} = \frac{(s_1^2/n_1 + s_2^2/n_2)^2}{(s_1^2/n_1)^2/(n_1-1) + (s_2^2/n_2)^2/(n_2-1)}.$$

which often approximates to (n_1+n_2-2). If this p-value is smaller than a desired *significance level* (say α=0.05) or the highest level of risk (probability) we are willing to take to conclude that there is a treatment effect when in fact there is none (Type I error), then we can reject the null hypotheses.

After demonstrating whether the treatment group has a significantly higher mean than the control group, the next question usually is what is the *effect size* (ES) or the magnitude of the treatment effect relative to the control group? We can estimate the ES by conducting regression analysis with performance scores as the outcome variable (y) and a dummy coded treatment variable as the predictor variable (x) in a two-variable GLM. The regression coefficient of the treatment variable (β_1), which is also the slope of the regression line ($\beta_1 = \Delta y/\Delta x$), is an estimate of the effect size. In the above example, since x is a dummy variable with two values (0 and 1), Δx = 1–0 = 1, and hence the effect size or β_1 is simply the difference between treatment and control means ($\Delta y = y_1 - y_2$).

Factorial Designs

Extending from the previous example, let us say that the effect of the special curriculum (treatment) relative to traditional curriculum (control) depends on the amount of instructional time (3 or 6 hours/week). Now, we have a 2 x 2 factorial design, with the two factors being curriculum type (special versus traditional) and instructional type (3 or 6 hours/week). Such a design not only helps us estimate the independent effect of each factor, called *main effects*, but also the joint effect of both factors, called the *interaction effect*. The generalized linear model for this two-way factorial design is designated as follows:

$$y = \beta_0 + \beta_1 x_1 + \beta_2 x_2 + \beta_3 x_1 x_2 + \varepsilon$$

where y represents students' post-treatment performance scores, x_1 is the treatment (special versus traditional curriculum), x_2 is instructional time (3 or 6 hours/week). Note that both x_1 and x_2 are dummy variables, and although x_2 looks like a ratio-scale variable (3 or 6), it actually represents two categories in the factorial design. Regression coefficients β_1 and β_2 provide effect size estimates for the main effects and β_3 for the interaction effect. Alternatively, the same factorial model can be analyzed using a two-way ANOVA analysis. Regression analysis involving multiple predictor variables is sometimes called multiple regression, which is different from multivariate regression that uses multiple outcome variables.

A note on interpreting interaction effects. If β_3 is significant, it implies that the effect of the treatment (curriculum type) on student performance depends on instructional time. In this case, we cannot meaningfully interpret the independent effect of the treatment (β_1) or of instructional time (β_2), because the two effects cannot be isolated from each other. Main effects are interpretable only when the interaction effect is non-significant.

Covariates can be included in factorial designs as new variables, with new regression coefficients (e.g., β_4). Covariates can be measured using interval or ratio scaled measures, even when the predictors of interest are designated as dummy variables. Interpretation of covariates also follows the same rules as that of any other predictor variable.

Other Quantitative Analysis

There are many other useful inferential statistical techniques, based on variations in the GLM, that are briefly mentioned here. Interested readers are referred to advanced text books or statistics courses for more information on these techniques:

- *Factor analysis* is a data reduction technique that is used to statistically aggregate a large number of observed measures (items) into a smaller set of unobserved (latent) variables called factors based on their underlying bivariate correlation patterns. This technique is widely used for assessment of convergent and discriminant validity in multi-item measurement scales in social science research.
- *Discriminant analysis* is a classificatory technique that aims to place a given observation in one of several nominal categories based on a linear combination of predictor variables. The technique is similar to multiple regression, except that the dependent variable is nominal. It is popular in marketing applications, such as for classifying customers or products into categories based on salient attributes as identified from large-scale surveys.
- *Logistic regression* (or logit model) is a GLM in which the outcome variable is binary (0 or 1) and is presumed to follow a logistic distribution, and the goal of the regression analysis is to predict the probability of the successful outcome by fitting data into a logistic curve. An example is predicting the probability of heart attack within a specific period, based on predictors such as age, body mass index, exercise regimen, and so forth. Logistic regression is extremely popular in the medical sciences. Effect size estimation is based on an "odds ratio," representing the odds of an event occurring in one group versus the other.
- *Probit regression (or* probit model) is a GLM in which the outcome variable can vary between 0 and 1 (or can assume discrete values 0 and 1) and is presumed to follow a standard normal distribution, and the goal of the regression is to predict the probability of each outcome. This is a popular technique for predictive analysis in the actuarial science, financial services, insurance, and other industries for applications such as credit scoring based on a person's credit rating, salary, debt and other information from her loan application. Probit and logit regression tend to demonstrate similar regression coefficients in comparable applications (binary outcomes); however the logit model is easier to compute and interpret.
- *Path analysis* is a multivariate GLM technique for analyzing directional relationships among a set of variables. It allows for examination of complex nomological models where the dependent variable in one equation is the independent variable in another equation, and is widely used in contemporary social science research.
- *Time series analysis* is a technique for analyzing time series data, or variables that continually changes with time. Examples of applications include forecasting stock market fluctuations and urban crime rates. This technique is popular in econometrics, mathematical finance, and signal processing. Special techniques are used to correct for auto-correlation, or correlation within values of the same variable across time.

Chapter 16

Research Ethics

Ethics is defined by Webster's dictionary as conformance to the standards of conduct of a given profession or group. Such standards are often defined at a disciplinary level though a professional code of conduct, and sometimes enforced by university committees called even Institutional Review Board. Even if not explicitly specified, scientists are still expected to be aware of and abide by general agreements shared by the scientific community on what constitutes acceptable and non-acceptable behaviors in the professional conduct of science. For instance, scientists should not manipulate their data collection, analysis, and interpretation procedures in a way that contradicts the principles of science or the scientific method or advances their personal agenda.

Why is research ethics important? Because, science has often been manipulated in unethical ways by people and organizations to advance their private agenda and engaging in activities that are contrary to the norms of scientific conduct. A classic example is pharmaceutical giant Merck's drug trials of Vioxx, where the company hid the fatal side-effects of the drug from the scientific community, resulting in 3468 deaths of Vioxx recipients, mostly from cardiac arrest. In 2010, the company agreed to a $4.85 billion settlement and appointed two independent committees and a chief medical officer to monitor the safety of its drug development process. Merck's conduct was unethical and violation the scientific principles of data collection, analysis, and interpretation.

Ethics is the moral distinction between right and wrong, and what is unethical may not necessarily be illegal. If a scientist's conduct falls within the gray zone between ethics and law, she may not be culpable in the eyes of the law, but may still be ostracized in her professional community, face severe damage to professional reputation, and may even lose her job on grounds of professional misconduct. These ethical norms may vary from one society to another, and here, we refer to ethical standards as applied to scientific research in Western countries.

Ethical Principles in Scientific Research

Some of the expected tenets of ethical behavior that are widely accepted within the scientific community are as follows.

Voluntary participation and harmlessness. Subjects in a research project must be aware that their participation in the study is voluntary, that they have the freedom to withdraw from the study at any time without any unfavorable consequences, and they are not harmed as a result of their participation or non-participation in the project. The most flagrant violations of the voluntary participation principle are probably forced medical experiments conducted by Nazi researchers on prisoners of war during World War II, as documented in the post-War Nuremberg Trials (these experiments also originated the term "crimes against humanity"). Less known violations include the Tuskegee syphilis experiments conducted by the U.S. Public Health Service during 1932-1972, in which nearly 400 impoverished African-American men suffering from syphilis were denied treatment even after penicillin was accepted as an effective treatment of syphilis, and subjects were presented with false treatments such as spinal taps as cures for syphilis. Even if subjects face no mortal threat, they should not be subjected to personal agony as a result of their participation. In 1971, psychologist Philip Zambardo created the Stanford Prison Experiment, where Stanford students recruited as subjects were randomly assigned to roles such as prisoners or guards. When it became evident that student prisoners were suffering psychological damage as a result of their mock incarceration and student guards were exhibiting sadism that would later challenge their own self-image, the experiment was terminated.

Today, if an instructor asks her students to fill out a questionnaire and informs them that their participation is voluntary, students must not fear that their non-participation may hurt their grade in class in any way. For instance, it in unethical to provide bonus points for participation and no bonus points for non-participations, because it places non-participants at a distinct disadvantage. To avoid such circumstances, the instructor may possibly provide an alternate task for non-participants so that they can recoup the bonus points without participating in the research study, or by providing bonus points to everyone irrespective of their participation or non-participation. Furthermore, all participants must receive and sign an **Informed Consent** form that clearly describes their right to not participate and right to withdraw, before their responses in the study can be recorded. In a medical study, this form must also specify any possible risks to subjects from their participation. For subjects under the age of 18, this form must be signed by their parent or legal guardian. Researchers must retain these informed consent forms for a period of time (often three years) after the completion of the data collection process in order comply with the norms of scientific conduct in their discipline or workplace.

Anonymity and confidentiality. To protect subjects' interests and future well-being, their identity must be protected in a scientific study. This is done using the dual principles of anonymity and confidentiality. **Anonymity** implies that the researcher or readers of the final research report or paper cannot identify a given response with a specific respondent. An example of anonymity in scientific research is a mail survey in which no identification numbers are used to track who is responding to the survey and who is not. In studies of deviant or undesirable behaviors, such as drug use or illegal music downloading by students, truthful responses may not be obtained if subjects are not assured of anonymity. Further, anonymity assures that subjects are insulated from law enforcement or other authorities who may have an interest in identifying and tracking such subjects in the future.

In some research designs such as face-to-face interviews, anonymity is not possible. In other designs, such as a longitudinal field survey, anonymity is not desirable because it prevents the researcher from matching responses from the same subject at different points in time for longitudinal analysis. Under such circumstances, subjects should be guaranteed **confidentiality**, in which the researcher can identify a person's responses, but promises not to divulge that person's identify in any report, paper, or public forum. Confidentiality is a weaker form of protection than anonymity, because social research data do not enjoy the "privileged communication" status in United State courts as do communication with priests or lawyers. For instance, two years after the Exxon Valdez supertanker spilled ten million barrels of crude oil near the port of Valdez in Alaska, the communities suffering economic and environmental damage commissioned a San Diego research firm to survey the affected households about personal and embarrassing details about increased psychological problems in their family. Because the cultural norms of many Native Americans made such public revelations particularly painful and difficult, respondents were assured confidentiality of their responses. When this evidence was presented to court, Exxon petitioned the court to subpoena the original survey questionnaires (with identifying information) in order to cross-examine respondents regarding their answers that they had given to interviewers under the protection of confidentiality, and was granted that request. Luckily, the Exxon Valdez case was settled before the victims were forced to testify in open court, but the potential for similar violations of confidentiality still remains.

In one extreme case, Rick Scarce, a graduate student at Washington State University, conducted participant observation studies of animal rights activists, and chronicled his findings in a 1990 book called *Ecowarriors: Understanding the Radical Environmental Movement*. In 1993, Scarce was called before a grand jury to identify the activists he studied. The researcher refused to answer grand jury questions, in keeping with his ethical obligations as a member of the American Sociological Association, and was forced to spend 159 days at Spokane County Jail. To protect themselves from travails similar to Rik Scarce, researchers should remove any identifying information from documents and data files as soon as they are no longer necessary. In 2002, the United States Department of Health and Human Services issued a "Certificate of Confidentiality" to protect participants in research project from police and other authorities. Not all research projects qualify for this protection, but this can provide an important support for protecting participant confidentiality in many cases.

Disclosure. Usually, researchers have an obligation to provide some information about their study to potential subjects before data collection to help them decide whether or not they wish to participate in the study. For instance, who is conducting the study, for what purpose, what outcomes are expected, and who will benefit from the results. However, in some cases, disclosing such information may potentially bias subjects' responses. For instance, if the purpose of a study is to examine to what extent subjects will abandon their own views to conform with "groupthink" and they participate in an experiment where they listen to others' opinions on a topic before voicing their own, then disclosing the study's purpose before the experiment will likely sensitize subjects to the treatment. Under such circumstances, even if the study's purpose cannot be revealed before the study, it should be revealed in a debriefing session immediately following the data collection process, with a list of potential riska or harm borne by the participant during the experiment.

Analysis and reporting. Researchers also have ethical obligations to the scientific community on how data is analyzed and reported in their study. Unexpected or negative findings should be fully disclosed, even if they cast some doubt on the research design or the findings. Similarly, many interesting relationships are discovered after a study is completed, by chance or data mining. It is unethical to present such findings as the product of deliberate design. In other words, hypotheses should not be designed in positivist research after the fact based on the results of data analysis, because the role of data in such research is to test hypotheses, and not build them. It is also unethical to "carve" their data into different segments to prove or disprove their hypotheses of interest, or to generate multiple papers claiming different data sets. Misrepresenting questionable claims as valid based on partial, incomplete, or improper data analysis is also dishonest. Science progresses through openness and honesty, and researchers can best serve science and the scientific community by fully disclosing the problems with their research, so that they can save other researchers from similar problems.

Institutional Review Boards

Research ethics in studies involving human subjects is governed in the United States by federal law. Any agency, such as a university or a hospital, that wants to apply for federal funding to support its research projects must establish that it is in compliance with federal laws governing the rights and protection of human subjects. This process is overseen by a panel of experts in that agency called an **Institutional Review Board** (IRB). The IRB reviews all research proposal involving human subjects to ensure that the principles of voluntary participation, harmlessness, anonymity, confidentiality, and so forth are preserved, and that the risks posed to human subjects are minimal. Even though the federal laws apply specifically for federally funded projects, the same standards and procedures are also applied to non-funded or even student projects.

The IRB approval process require completing a structured application providing complete information about the research project, the researchers (principal investigators), and details on how the subjects' rights will be protected. Additional documentation such as the Informed Consent form, research questionnaire or interview protocol may be needed. The researchers must also demonstrate that they are familiar with the principles of ethical research by providing certification of their participation in an research ethics course. Data collection can commence only after the project is cleared by the IRB review committee.

Professional Code of Ethics

Most professional associations of researchers have established and published formal codes of conduct describing what constitute acceptable and unacceptable professional behavior of their member researchers. As an example, the summarized code of conduct for the Association of Information Systems (AIS), the global professional association of researchers in the information systems discipline, is summarized in Table 16.1 (the complete code of conduct is available online at http://home.aisnet.org/ displaycommon.cfm?an=1&subarticlenbr=15). Similar codes of ethics are also available for other disciplines.

The AIS code of conduct groups ethical violations in two categories. Category I includes serious transgressions such as plagiarism and falsification of data, research procedures, or data analysis, which may lead to expulsion from the association, dismissal from employment, legal action, and fatal damage to professional reputation. Category 2 includes less serious transgression such as not respecting the rights of research subjects, misrepresenting the originality of research projects, and using data published by others without acknowledgement, which may lead to damage to professional reputation, sanctions from journals, and so forth. The code also provides guidance on good research behaviors, what to do when ethical transgressions are detected (for both the transgressor and the victim), and the process to be followed by AIS in dealing with ethical violation cases. Though codes of ethics such as this have not completely eliminated unethical behavior, they have certainly helped clarify the boundaries of ethical behavior in the scientific community and reduced instances of ethical transgressions.

CATEGORY ONE: Codes in this category must ALWAYS be adhered to and disregard for them constitutes a serious ethical breach. Serious breaches can result in your expulsion from academic associations, dismissal from your employment, legal action against you, and potentially fatal damage to your academic reputation. 1. Do not plagiarize. 2. Do not fabricate or falsify data, research procedures, or data analysis.
CATEGORY TWO: Codes in this category are recommended ethical behavior. Flagrant disregard of these or other kinds of professional etiquette, while less serious, can result in damage to your reputation, editorial sanctions, professional embarrassment, legal action, and the ill will of your colleagues. 3. Respect the rights of research subjects, particularly their rights to information privacy, and to being informed about the nature of the research and the types of activities in which they will be asked to engage. 4. Do not make misrepresentations to editors and conference program chairs about the originality of papers you submit to them. 5. Do not abuse the authority and responsibility you have been given as an editor, reviewer or supervisor, and ensure that personal relationships do not interfere with your judgement. 6. Declare any material conflict of interest that might interfere with your ability to be objective and impartial when reviewing submissions, grant applications, software, or undertaking work from outside sources. 7. Do not take or use published data of others without acknowledgement, or unpublished data without both permission and acknowledgement. 8. Acknowledge the substantive contributions of all research participants, whether colleagues or students, according to their intellectual contribution. 9. Do not use other people's unpublished writings, information, ideas, concepts or data that you may see as a result of processes such as peer review without permission of the author. 10. Use archival material only in accordance with the rules of the archival source.
ADVICE: Some suggestions on how to protect yourself from authorship disputes, mis-steps, mistakes, and even legal action. 1. Keep the documentation and data necessary to validate your original authorship for each scholarly work with which you are connected. 2. Do not republish old ideas of your own as if they were a new intellectual contribution. 3. Settle data set ownership issues before data compilation. 4. Consult appropriate colleagues if in doubt.

Table 16.1. Code of ethics for the Association of Information Systems

An Ethical Controversy

Robert Allen "Laud" Humphreys is an American sociologist and author, who is best known for his Ph.D. dissertation, *Tearoom Trade*, published in 1970. This book is an ethnographic account of anonymous male homosexual encounters in public toilets in parks – a practice known as "tea-rooming" in U.S. gay slang. Humphreys was intrigued by the fact that the majority of participants in tearoom activities were outwardly heterosexual men, who lived otherwise conventional family lives in their communities. However, it was important to them to preserve their anonymity during tearoom visits.

Typically, tearoom encounters involved three people – the two males engaging in a sexual act and a lookout person called a "watchqueen." The job of the watchqueen was to alert the two participating males for police or other people, while deriving pleasure from watching the action as a voyeur. Because it was not otherwise possible to reach these subjects, Humphreys showed up at public toilets, masquerading as a watchqueen. As a participant observer, Humphreys was able to conduct field observations for his dissertation, as he normally would in a study of political protests or any other sociological phenomenon.

Humphreys needed more information on the participants. But because participants were unwilling to be interviewed in the field or disclose personal identities, Humphreys wrote down the license plate numbers of the participants' cars, wherever possible, and tracked down their names and addresses from public databases. Then he visited these men at their homes, disguising himself to avoid recognition and announcing that he was conducting a survey, and collected personal data that was not otherwise available.

Humphreys' research generated considerable controversy within the scientific community. Many critics said that he should not have invaded others' right to privacy in the name of science, others were worried about his deceitful behavior in leading participants to believe that he was only a watchqueen, when he clearly had ulterior motives. Even those who considered observing tearoom activity to be acceptable because the participants used public facilities, thought that the follow-up interview survey in participants' homes under false pretenses was unethical, because of the way he obtained their home addresses and because he did not seek informed consent. A few researchers justified Humphrey's approach saying that this was an important sociological phenomenon worth investigating, that there was no other way to collect this data, and that the deceit was harmless, since Humphreys did not disclose his subjects' identities to anyone. This controversy was never resolved, and it is still hotly debated in classes and forums on research ethics.

Appendix:

Sample Syllabus for a Doctoral Course

QMB 7565: Introduction to Research Methods
University of South Florida
College of Business
Fall Semester 2011

Objectives:

The purpose of this course is to introduce doctoral students to the process of conducting academic research. We will learn about how to think and act like a researcher in conceptualizing, designing, executing, and evaluating "scientific" research projects. Part of the class will require you to design and write an independent research proposal (with your professor's help). In addition, the finals exam will prepare you for the research methods section of your Ph.D. comprehensives exam.

Structure:

This class is designed in a seminar format. This heart of any Ph.D. seminar is discussion and analysis of assigned readings. To do that, you must read all assigned readings before class, think about these issues throughout the semester, debate these issues with your classmates, and synthesize these issues mentally to develop yourself as a researcher. Please note that if you do not come to class fully prepared, you will be totally lost and will not learn anything from the class.

Being a doctoral seminar, this course will entail: (1) a much higher workload than any Masters-level course you have encountered thus far, (2) a heavy dose of boring readings, and (3) a substantial amount of critical and often frustrating thinking. This is not an easy class and you will not receive an easy grade. If you are challenged by the demanding nature of this class, you should drop this class and drop out of the Ph.D. program.

Books and Materials:

Bhattacherjee, A. *Social Science Research: Principles, Methods, and Practice*, Ver 2.0, 2011, Free download from Blackboard (my.usf.edu).

Kuhn, T. J. *The Structure of Scientific Revolutions*, University of Chicago Press, Chicago, 1996. This entire book must be read prior to Week 4. This $7.50 paperback (Amazon price) looks deceptively small, but is actually a heavy drill, so start reading it right away.

Papers: Download from Blackboard (my.usf.edu). See Class Schedule.

Grading:

Grade components:	Finals exam	35 points
	Research proposal	35 points
	Paper review	10 points
	Class participation	20 points

Grading scale: A+: 97-100; A: 92-97; A-: 90-92; B+: 88-90; B: 82-88; B-: 80-82; C+: 78-80; C: 70-78

Finals Exam:

This is a 3-hour, comprehensive exam, in which you will be asked 3-4 multi-part essay-type questions, similar to the ones in your Ph.D. comprehensive exams. Sample exam questions can be downloaded from Blackboard. You can either hand-write your answers on type them a computer. Your answers should demonstrate (a) synthesis of materials covered in class and (b) your own critical analysis of these materials, and not a mere regurgitation of the papers or the professor's comments. You will be graded solely on quality of your answer and not on how many pages you write. You will lose credit if your answer is irrelevant (of the rambling type), does not contain enough details (too general or vague), or demonstrate just a superficial level of understanding of the materials discussed in class. Exam is open-book/open-notes, but please be forewarned that having all the materials in front of you will not help, if you didn't bother to prepare for the exams ahead of time. Given the avalanche of topics and materials covered in class, cramming the week before the exams will be too little and too late.

Research Proposal:

You will apply your learning about the research process into a research proposal that is due at the end of the semester. This proposal will be 10-12 pages single-spaced, excluding references and appendices. You can select any research problem of your choice for this proposal, but preferably in the positivist tradition of inquiry. However, the proposed research must be original (i.e., not something you are doing for another class or another professor), must examine a real problem (i.e., not a hypothetical or "toy" problem), and must be of at least a conference-level quality. Your proposal must include five sections: (1) research problem and significance, (2) literature review, (3) theory and hypotheses, (4) research methods, and (5) research plan. Data collection or analysis is not necessary. But if you want to do an interpretive research project instead, then some data collection and analysis will be needed (and the above structure will also change – talk to me about these changes). Project deliverables are due throughout the course of the semester, as we cover corresponding topics in class. This will allow me to give you early feedback and correct problems well before the final due date. Note that intermediate deliverables are not graded, only your final proposal will be graded. You will also present your final proposal during the last week of class in a 15-20 minute formal presentation (plus a 5 minute question/answer session), as you would typically do at an academic conference.

Paper Review:

A key component of an academic career is critically evaluating others' research. During the second half of the semester, you'll be asked to write a formal review (critique) of a paper submitted for publication to a leading business journal. To help you prepare, I'll discuss in class how to write reviews, give you a framework for review, and have you do one practice round of review in class for an actual journal submission. After you write your review, I will give you actual reviews by anonymous reviewers/editors for this paper so that you can compare your own review with that of professional reviewers, and see what you missed. The graded paper review is take-home, and due one week before the finals. You may need to do some background research or read some additional papers before writing your review report, but you must not collaborate or discuss the paper with anyone in or out of class.

Class Participation:

Each paper discussed in class will be assigned a "lead discussant," who will be responsible for (1) preparing a one-page synopsis of those papers (in a structured format) and (2) leading class discussion on that paper. You will receive full class participation grade if you turn in synopsis of all assigned papers on time and do a reasonable job in class discussion. Please see the sample synopsis to get an idea of how to structure these synopses. These synopses can be useful study aids for the exams, but only if you do a thorough job with them. Bring enough copies of the synopses to share with the rest of the class. I'll give you immediate feedback on your synopsis as we discuss the paper in class, so that you can try to improve your synopsis next time. If you have to miss a class or would like to do a different paper, you can exchange your assigned paper with a classmate. However, if there is a miscommunication and no synopsis is available for a paper on the discussion date, the student originally assigned that paper will be penalized in his/her class participation grade. Irrespective of synopsis assignment, everyone is expected to attend all classes, read all papers, and contribute to all class discussions.

Class Policies:

Attendance: I do not take formal attendance, but I do keep track of who is coming to class and who is not. If you think you will miss more than one week of class during the semester, you should drop this class.

Academic honesty: Plagiarism in any form is banned and will result in a straight FF grade. Please refer to USF's academic honesty policy in your student handbook.

Disability: Students requiring disability accommodations should notify me within the first two weeks of class, with a letter from the Student Disability Services Office.

Cell phones: Cell phones must be turned OFF during the duration of the class, but you can use them during the breaks.

Class Schedule:

Week 1: Introduction to Research

Syllabus & Introductions.
Chapter 1.
The Research & Publication Process.

Week 2: Thinking Like a Researcher

Chapter 2.
A Tale of Two Papers:
- Fichman, R.G. and Kemerer, C.F., "The Illusory Diffusion of Innovation: An Examination of Assimilation Gaps," Information Systems Research (10:3), September 1999, pp. 255-275.
- Williams, L.; Kessler, R.R.; Cunningham, W.; and Jeffries, R., "Strengthening the Case for Pair Programming," IEEE Software, July/August 2000, pp. 19-25.

Critical Thinking:
- Dialog on Leadership, "Awareness is the First Critical Thing," A Conversation with Wanda Orlikowski, 1999.

Research Ethics:
- AIS Code of Conduct: http://home.aisnet.org/displaycommon.cfm?an=1&subarticlenbr=13
- IRB Process: http://www.research.usf.edu/cs/irb_forms.htm

Week 3: Literature Review & Meta-Analysis

Lit review:
- Fichman, R.G., "Information Technology Diffusion: A Review of Empirical Research," Proceedings of the Thirteenth International Conference on Information Systems, Dallas, 1992, 195-206.
- Alavi, M. and Leidner, D.E., "Knowledge Management and Knowledge Management Systems," MIS Quarterly (25:2), March 2001, pp. 107-136.

Meta-analysis:
- King, W.R, and He, J., "Understanding the Role and Methods of Meta-Analysis in IS Research," Communications of the AIS (16), 2005, pp. 665.686.
- Henard, D.H. and Szymanski, D.M., "Why Some New Products are More Successful Than Others," Journal of Marketing Research (38), August 2001, pp. 362-375.

Due: Research Proposal: Research Problem and its Significance.

Week 4: Philosophy of Science

Chapter 3.
Paradigms in scientific inquiry:
- Kuhn, T., The Structure of Scientific Revolutions, University of Chicago Press, 1996 (entire book).
  [See Dr. Pajares' notes ONLY if you completely lost]

Social science paradigms:
- Krugman, P., "How Did Economists Get It So Wrong," New York Times, Sept 6, 2009.
- Gioia, D.A. and Pitre, E., "Multiparadigm Perspectives on Theory Building," Academy of Management Review (15:4), 1990, pp. 584-602.

Week 5: Theories in Organizational Research

Chapter 4.
Why theory:
- Steinfield, C.W. and Fulk, J., "The Theory Imperative," in Organizations and Communications Technology, Janet Fulk and Charles W. Steinfield (eds.), Sage Publications, Newbury Park, CA, 1990.

Agency theory:
- Eisenhardt, K.M., "Agency Theory: An Assessment and Review," Academy of Management Review (14:1), 1989, pp. 57-74.

Transaction cost theory:
- Williamson, O.E., "The Economics of Organization: The Transaction Cost Approach," American Journal of Sociology (87:3), 1981, pp. 548-577.

Week 6: Organizational Theories - Continued

Resource-based theory:
- Barney, J.B., "Firm Resources and Sustained Competitive Advantage," Journal of Management (17:1), 1991, pp. 99-120.
- Priem, R.L. and Butler, J.E., "Is the Resource-Based 'View' a Useful Perspective for Strategic Management Research?", Academy of Management Review (26:1), 2001, pp. 22-40.

Dynamic capability theory:
- Teece, D.J.; Pisano, G.; and Shuen, A., "Dynamic Capabilities and Strategic Management," Strategic Management Journal (18:7), 1997, 509-533.

Due: Research Proposal: Literature Review (with prior sections, modified if needed).

Week 7: Measurement and Validity

Chapters 6 and 7.
Scale validity and unidimensionality:
- Straub, D.W., "Validating Instruments in MIS Research," MIS Quarterly (13:2), June 1989, pp. 146-169.
- MacKenzie, S. B., Podsakoff, P. M., and Podsakoff, N. P., "Construct Measurement and Validation Procedures in MIS and Behavioral Research: Integrating New and Existing Techniques," MIS Quarterly (35:2), 2011, pp. 293-334.

Due: Research Proposal: Theory & Hypotheses (with prior sections, modified as needed).

Week 8: Survey Research

Chapters 5, 8, and 9.
Field survey exemplar:
- Tsai, W., "Knowledge Transfer in Intraorganizational Networks: Effects of Network Position and Absorptive Capacity on Business Unit Innovation and Performance," Academy of Management Review, 2001.

Biases in survey research:
- Malhotra, N. K, Kim, S. S., and Patil, A., "Common Method Variance in IS Research: A Comparison of Alternative Approaches and a Reanalysis of Past Research," Management Science (52:12), 2006, pp. 1865-1883.

Week 9: Experimental & Quasi-Experimental Research

Chapter 10.
Field experiment examplar:

- Hunton, J.E. and McEwen, R.A., "An Assessment of the Relation Between Analysts' Earnings Forecast Accuracy, Motivational Incentives, and Cognitive Information Search Strategy," The Accounting Review (72:4), October 1997, pp. 497-515.

Problems with experimental research:

- Jarvenpaa, S. L.; Dickson, G. W.; and DeSanctis, G., "Methodological Issues in Experimental IS Research: Experiences and Recommendations," MIS Quarterly, June 1985, pp. 141-156.

Week 10: Reviewing Research

Writing paper reviews:

- Lee, A.S., "Reviewing a Manuscript for Publication," Journal of Operations Management (13:1), July 1995, pp.87-92.
- Agarwal, R; Echambadi, R; Franco, A.M.; and Sarkar, M.B., "Reap Rewards: Maximizing Benefits from Reviewer Comments," Academy of Management Journal (49:2), 2006, pp. 191-196.

Why papers are rejected:

- Daft, R.L., "Why I Recommended that Your Manuscript be Rejected and What You Can Do About It," in L.L. Cummings & P.J. Frost (eds.), Publishing in the Organizational Sciences, 2nd ed., 1995, pp. 164-182.

Formal review:

- Write a formal review of the following paper based on the guidelines prescribed in the above articles. Use this review template to structure your review: Anonymous, Paper submitted to MIS Quarterly, 2004.
  Actual comments from three reviewers, AE, and SE will be e-mailed to you the day before (after you complete your review).

Paper review:

- One "mystery paper" for you to review and submit on Week 13. This is part of your course grade, hence please plan to spend adequate time and effort on this review.

Week 11: Case Research

Chapter 11.
Conducting case research:

- Benbasat, I.; Goldstein, D.K.; and Mead, M., "The Case Research Strategy in Studies of Information Systems," MIS Quarterly, September 1987, pp. 369-386.

Case research exemplars:

- Beaudry, A. and Pinsonneault, A., "Understanding User Responses to Information Technology: A Coping Model of User Adaptation," MIS Quarterly (29:3), September 2005, pp. 493-524.
- Eisenhardt, K.M., "Making Fast Strategic Decisions In High-Velocity Environments," Academy of Management Journal (32:3), 1989, pp. 543-577.

Positivist versus Interpretive Analysis:

- Trauth, E.M. and Jessup, L.M., "Understanding Computer-Mediated Discussions: Positivist and Interpretive Analyses of Group Support System Use," MIS Quarterly (24:1), March 2000, pp. 43-79.

Due: Research Proposal: Research Methods (with all prior sections).

Week 12: Interpretive Research

Chapters 12 and 13.
Demo: Content Analysis using Grounded Theory.
Qualitative research:

- Shah, S.K. and Corley, K.G., "Building Better Theory by Bridging the Quantitative-Qualitative Divide," Journal of Management Studies (48:3), December 2006, pp. 1821-1835.

Action research:

- Kohli, R. and Kettinger, W., "Informating the Clan: Controlling Physician Costs and Outcomes," MIS Quarterly (28:3), September 2004, pp. 1-32.

Ethnography:

- Barley, S.R., "Technicians in the Workplace: Ethnographic Evidence for Bringing Work into Organization Studies," Administrative Science Quarterly (41), 1996, pp. 404-411.

Week 13: Miscellaneous Methods

Chapter 16.
Demo:

- Statistical analysis using SPSS.

Secondary data analysis:

- Chaney, P.K. and Philipich, K.L., "Shredded Reputation: The Cost of Audit Failure," Journal of Accounting Research (40:4), September 2002, pp. 1221-1245.

Analytic Modeling:

- Bayus, B.L., Jain, S., and Rao, A.G, "Truth or Consequences: An Analysis of Vaporware and New Product Announcements," Journal of Marketing Research (38), February 2001, pp. 3-13.

Due: Article Review.

Week 14: Finals

Comprehensive in-class 3-hour open-book exam. See Sample Exam Questions.

Week 15: Student Presentations

20-minute presentation plus 5-minute Q & A.
Due: Final Research Proposal.

78267072R00087

Made in the USA
Columbia, SC
05 October 2017